D0930799

PACKAGING DESIGN 3

PACKAGING DESIGN 3

The Best of American and International Packaging Designs

by Cristina Gabetti and
the Editors of *ID* Magazine

PBC INTERNATIONAL, INC. • New York

Distributor to the book trade in the United States:

Rizzoli International Publications, Inc.
597 Fifth Avenue
New York, NY 10017

Distributor to the art trade in the United States:

Letraset USA
40 Eisenhower Drive
Paramus, NJ 07653

Distributor in Canada:

Letraset Canada Limited
555 Alden Road
Markham, Ontario L3R 3L5, Canada

Distributed throughout the rest of the world by:

Hearst Books International
1790 Broadway
New York, NY 10019

Library of Congress Cataloging-in-Publication Data

Gabetti, Cristina.
Packaging designs 3.

Includes Index.
1. Packaging--Design and construction.
I. ID (New York, N.Y.) II. Title. III. Title:
Packaging designs three.
TS195.2.G33 1987 688.8 87-1745
ISBN 0-86636-019-0

Printed and Bound by Arti Grafiche Motta Milan, Italy

10 9 8 7 6 5 4 3 2 1

STAFF

Publisher	**Herb Taylor**
Project Director	**Cora Sibal Taylor**
Executive Editor	**Virginia Christensen**
Editor	**Joanne Bolnick**
Art Director	**Richard Liu**
Art/Prod. Coordinator	**Jeanette Forman**
Artist	**Daniel Kouw**

CONTENTS

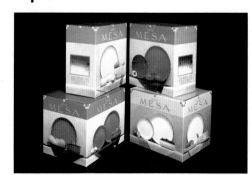

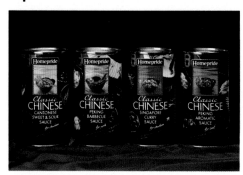

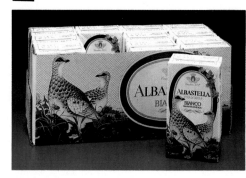

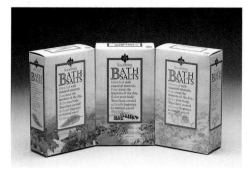

Foreword

This is the third volume of *Packaging Design* in a series begun in 1984 by PBC International. *Packaging Design 3* marks the beginning of annual publication, a step we are taking to keep up with the fast-evolving packaging design discipline. A new dimension, however, has been added by Cristina Gabetti, whose experience in packaged goods advertising and marketing, both in the U.S. and in Italy, expands the series to include extraordinary designs from Europe. The expertise and resources brought to this volume by Ms. Gabetti and by the editors of *ID* magazine make *Packaging Design 3* by far the most complete and up-to-date collection of the best in the field ever published. Not only did Ms. Gabetti tap into the European scene, but she was able to draw on the *ID Annual Design Review* entries as well as *ID's* regular contacts in the U.S.

Designing packaging is a tough assignment, for ideal solutions must meet so many different criteria and operate in all sorts of constraints. All of a designer's resources and skills must be tapped if he or she is to arrive at the best, most elegant design solution, although reaching the final design is also a function of the sensitivity of the client, who almost always tends to be conservative and often must be pulled and pushed into making the best design choice. It is the intention of *Packaging Design 3* to serve as a resource of ideas and information that will help both design professionals and clients bring about effective, creative design solutions that achieve design excellence.

R.McA.
Publisher, *ID* Magazine

INTRODUCTION

Two trends make this third edition of *Packaging Design* stand apart from the previous two. First, there is "the disappearing retail clerk." Second, there is *Eurostyle*.

The old cliché that packaging is the silent salesman is now true in a literal way. The cliché, of course, meant that a package, say, on a supermarket shelf had to "sell" the consumer in just a few seconds. Now, because department stores and mass merchandisers have greatly reduced the number of clerks behind their counters, the package must aspire to higher ambitions; it must describe the product, educate the consumer about the product, and promote the product's benefits. This new requirement has led to the repackaging of thousands of products that were formerly displayed in shipping containers without graphic packaging. It has also forced designers to recall their most basic training in using a grid to organize relatively complex material for a package. And it has also meant that designer and copywriter must work together to shape the graphic message.

In addition, this merchandising need is related to the second major influence, Eurostyle, a catch-all word that essentially refers to the use of Swiss-developed techniques that result in a clean-lined, performance-inferring, and upscale look. There is an important practical side to Eurostyle, for it is the ideal grid-based technique for organizing the mass of information a consumer requires when there are no sales clerks to answer questions. Also related to Eurostyle is the trend toward more structure in line packaging, an orderly and systematic approach which takes its cue from larger corporate identity programs. This umbrella approach also permits the smooth introduction of related items and line extensions, and helps both the consumer and the salesperson understand, from the package alone, the complicated full line of products.

The requirement to display masses of information has led to the introduction of an entire new range of outer packaging materials, paper, and printing and construction techniques that permit optimum graphic flexibility for designers. It also has given designers larger dimensions to work with and has permitted the extensive use of high-quality product photography on hard goods packages, providing more opportunities to illustrate end-use and multi-use.

Other lesser trends are also apparent in current packaging. One has to do with the comeback of a traditional type of packaging—in the name of naturalness and old-fashioned imagery. In the nineteenth century, American packages were heavily decorated, gussied up jobs, with one material often substituting for another—faux metal, or faux wood—in the spirit of vinyl woodgrain. The decorated package is again being used—for gourmet foods and gift items, and just about anything called "old fashioned." Its comeback has been inspired by the general interest in naturalness, and as a result, one might find foods wrapped in a package decorated with a woody or burlap sack motif.

Another lesser trend has to do with the application of the seductiveness seen in cosmetic packaging to other products. Cosmetic and fragrance packaging is a unique craft that endeavors to sell beauty, love, mystery, romance, sophistication, wealth and sex—all commodities hard to come by at one time or another in one's life. The frankly seductive quality of this kind of package tests the imagination of designers. Now, the seductiveness of cosmetic packaging is being applied elsewhere to enhance the luxury factor, particularly in such gift categories as fine soft goods, accessories, and stationery, where super-high-quality packaging "justifies" much higher margins.

Whatever the current trends, one cannot escape the tough reality of packaging design. The designer can control the look of the package, but has absolutely no control over where it will be seen by the public. The decision, then, to stick with a familiar look (and its historical value) or opt for what might lead to greater impact through design always carries a great degree of uncertainty. What's more, since so many other factors influence the sales of a product, it is hard to objectively isolate what factors, in reality, have the greatest influence.

Market researchers continue to struggle with this issue, devising new ways to test the consumer's reaction to a packaging design. These range from simple show-and-tell sessions, to various "eye camera" techniques, to in-depth focus group research. All provide valuable guidance for designers; added to this is the experience of the marketer, the influence of the retail environment, and the competitive situation.

One important thing does seem clear: designers and marketers need not "design down" to consumers as much as they have in the past, for the consumer today is more design conscious and tends to respond positively to simplicity and orderliness. The design taste level of American consumers has improved in the last few years, and this is good for designers, for it leads to creating, in Raymond Loewy's words, the "most advanced yet acceptable" design, rather than cooking up a stew that we perceive the public wants.

Randolph McAusland
Publisher, *ID* Magazine

Chapter 1

Foods

- Frozen entrees
- Condiments
- Seafoods
- Cookies
- Dairy products
- Diet foods
- Health foods
- Baby foods
- Candies

Because food packaging represents such a significant, and thus important, portion of the packaging industry, a trend that emerges there can signal packaging changes yet to come to the broader industry. Currently, three trends, or themes, appear to dominate food packaging, which accounts for much of the exceptional design now in the marketplace.

The first trend has to do with the updating of existing packages. In many instances, packages that have, for years, used traditional graphics to tout product features are now reaching for a more contemporary, up-to-date look. This often involves introducing a fresh, clean design that is simpler and thus more striking than the earlier design. Even the ageless Dundee marmalade jar (p. 81), while retaining some traditional graphics, has a fresh new appearance that is cleaner and simpler.

The second trend involves a more systematic approach to designing food packages, which have always tended to feature higly promotional and somewhat disorganized graphics. This systematic approach is particularly evident in the packaging for full product lines in product categories such as juices, cereals, cookies, and frozen foods.

Most product line extensions use the same-shaped package or container and the same basic graphic design but distinguish between products through color coding and/or different product photographs. Products that employ both of these methods include Durkee sauce and gravy mixes (p. 16), Nestea Ice Teasers (p. 21), Lipton Long Grain and Wild Rice (p. 30), le Morbide Promesse (p. 33), Shake 'n Bake (p. 40), Fruit & Fibre (p. 41), and Henri Nestle chocolate (p. 45). Others, such as Moul-bie flour (p. 22) and Nestle chocolate bars (p. 74), use color coding, together with customized copy, to identify specific products within the line.

The third trend has to do with the degree to which the packaging makes a sales pitch. As is evident among the examples in this chapter, the tendency to put as much promotional information as possible on the outer package is a particularly American phenomenon. This contrasts dramatically with the more understated European approach apparent in the delightful designs for Chocolate Eggs Alemanga (p. 14) and Honnies nut pralines (p. 20) and the elegant packages for Buxted Chicken Kiev (p. 28) and Randstad Uitzendbureau chocolate (p. 42).

In creating the packaging for Naturell Yoghurt, a product of the Swedish Dairy Cooperative, designer Tom Hedqvist turned to the understated design approach evident in most European food packaging. Strong dark photographs of vibrant elders, seen against a white background, create a truly eye-catching package without the busyness often associated with American food packaging.

The package for Pillsbury's BEST refrigerated cookies, designed by the firm of Kornick Lindsay, illustrates two trends in current food packaging. The graphics for this product line are clean, simple, and contemporary—luscious product photography is set against a white background visually softened by light-colored rules. In addition, the basic package design is kept the same for all items in the product line, with color coding and product photographs used for identification.

Product: Chocolate Eggs Alemagna
Designer: Studio Vu Srl
Design Firm: Studio Vu Srl, Milano, Italy
Client: SIDALM S.p.A., Milano, Italy

Symbols of peace and beauty painted in bright, cheerful colors welcome the Easter season.

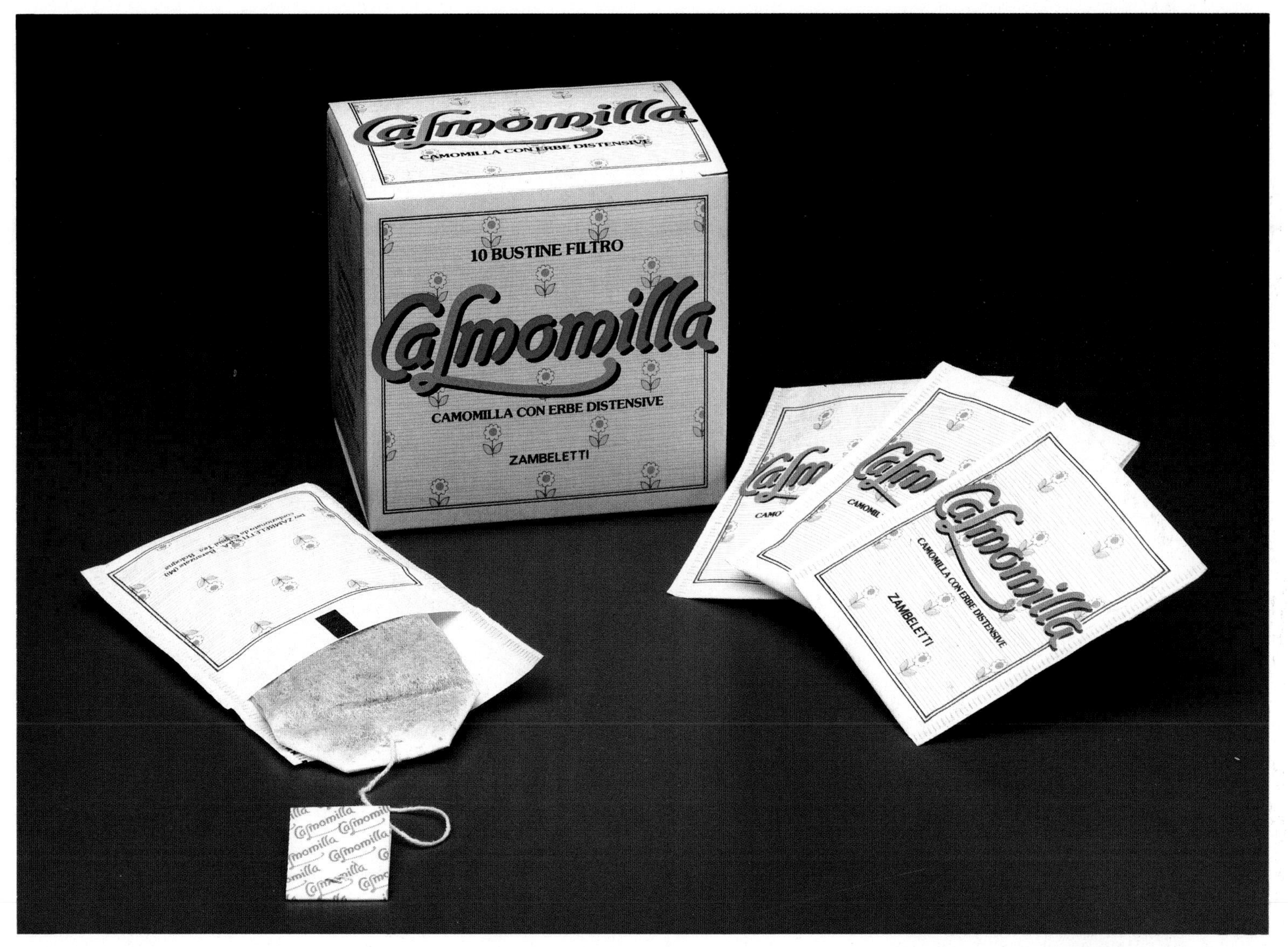

Product: Calmomilla
Designer: Gió Rossi
Design Firm: Image Plan International, Milano, Italy
Client: Zambelletti, DR.L., Italy

Little flowers on a yellow background label this soothing tea, which is a blend of chamomile and other calming herbs.

Product: County Line Natural Cheese
Designer: Dickens Design Group
Design Firm: Dickens Design Group, Chicago, IL
Client: Beatrice Cheese Company, New Berlin, WI

Total redesigning emphasizes County Line's new brandmark and quality-image packaging. New color coding helps consumers choose different cheese types.

Product: Durkee Sauce and Gravy Mixes
Designers: Sal LiPuma and John Rutig
Design Firm: Coleman, LiPuma, Segal & Morrill Inc., New York, NY
Client: Durkee Famous Foods Inc.

Simple, appetizing photography, a strong Durkee brand name image, and a distinct color-coding system were employed to make the packages of forty-five Durkee products stand out.

Product: Tasty Pastry Petites
Designers: Owen W. Coleman and Edward Morrill
Design Firm: Coleman, LiPuma, Segal & Morrill Inc., New York, NY
Client: Thomas J. Lipton Inc.

The new packaging for Pastry Petites, tailored for an assortment of flavors, creates a contemporary feeling with casual script logostyling for the brand name.

Product: Crabtree & Evelyn Cookies
Designer: Peter Windett & Associates
Design Firm: Peter Windett & Associates, London, England
Client: Crabtree & Evelyn, London, England

These fresh, hand-baked cookies are packaged in traditional, decorative English cookie tins designed with a touch of Christmas festivity.

Product: Sanka
Designers: Sal LiPuma, Abe Segal, and Cathy Sze-Tu
Design Firm: Coleman, LiPuma, Segal & Morrill Inc., New York, NY
Client: General Foods Corporation, White Plains, NY

This design represents a new upgraded, contemporary image for Sanka. The design theme is carried across all Sanka-brand package configurations.

Product: Barilla Egg Pasta
Designer: Michael Peters
Design Firm: Michael Peters and Partners, London, England
Client: Barilla G. & R. F.lli S.p.A., Italy

Because Barilla uses high quality ingredients, its egg pasta is positioned at the top of the Italian market. Its cellophane packaging allows consumers to see the quality and freshness of the product. Production costs are kept down because cellophane is less expensive than cardboard.

Product: Startine Spread
Designer: Shining
Design Firm: Shining, France
Client: Gervais Danone, France

Different spreads are packaged in similar reusable containers. Each flavor, like herb or bleu aux noix, is packaged in a color-coded container.

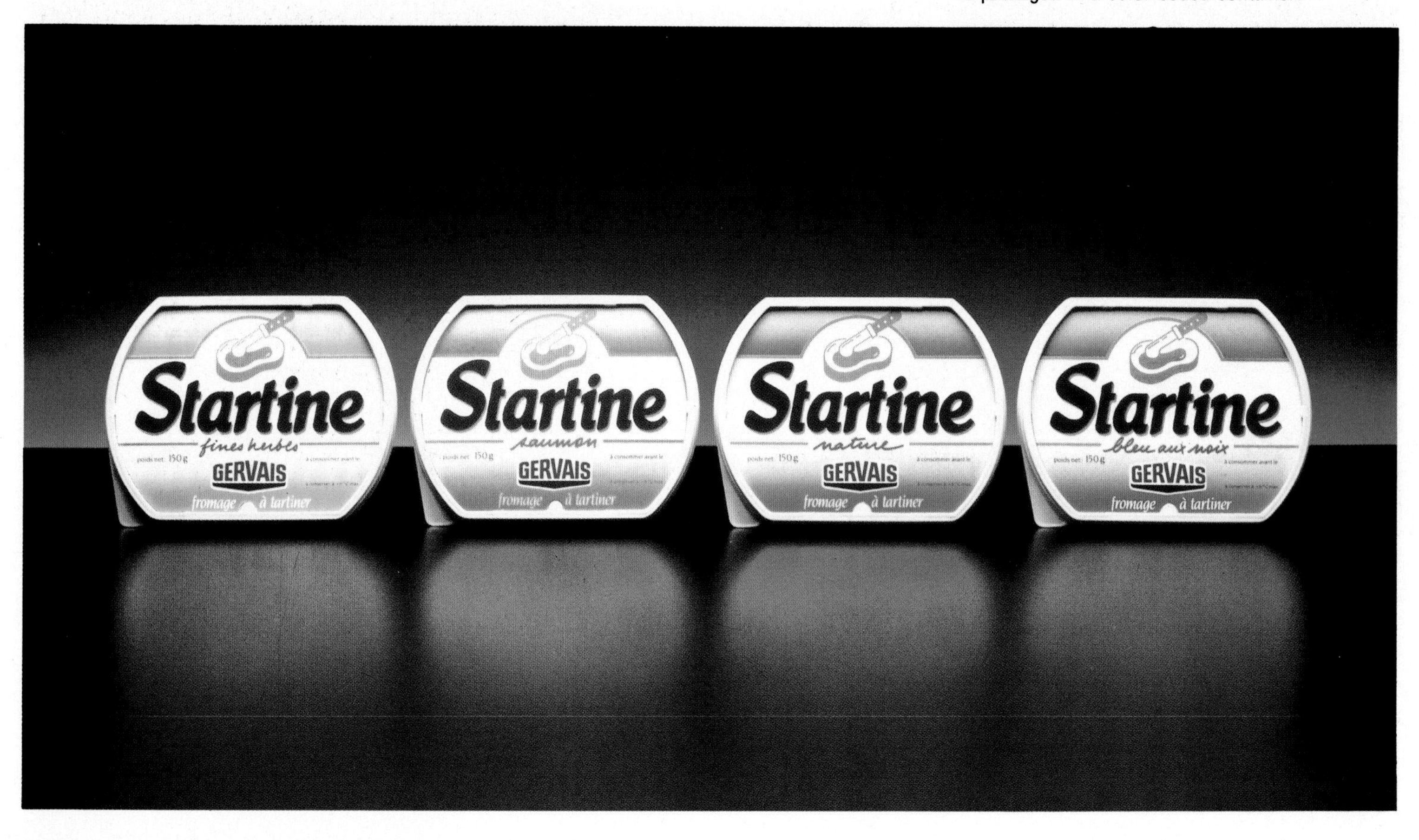

Product:	Barilla Pasta
Designer:	Michael Peters
Design Firm:	Michael Peters and Partners, London, England
Client:	Barilla G. & B. F.lli S.p.A., Italy

This kilogram package of Barilla's blue-label spaghetti is sold exclusively in Southern Italy where pasta consumption is highest. Cellophane wrapping caters to buyers who want to see the product and check the quality of what they are purchasing. This package improved sales for Barilla.

Product: Honnies Nut Pralines
Designer: Hotshop
Design Firm: Hotshop, Paris, France
Client: Bahlsen, France

The unique packaging shape is an interesting aspect of this product, as well as the illustration of a bear above the logo.

Product: Ice Cream Packaging
Designers: Ward M. Hooper and J.C. Chou
Design Firm: Coleman, LiPuma, Segal & Morrill Inc., New York, NY
Client: General Biscuit Brands Inc., Elizabeth, NJ

This design, developed for General Biscuit Brands, is for two ice cream cartons. The ice cream sandwich package uses a generic design that enables customers to imprint their own name.

Product: Fisher Nuts
Designers: Owen W. Coleman, Abe Segal, and John Rutig
Design Firm: Coleman, LiPuma, Segal & Morrill Inc., New York, NY
Client: Beatrice/Hunt-Wesson Inc., Fullerton, CA

The shelf visibility of the Fisher brand name has been improved by using appetizing photography to separate the brand name from the descriptive copy.

Product: Nestea Ice Teasers
Designers: Owen W. Coleman, Abe Segal, and John Rutig
Design Firm: Coleman, LiPuma, Segal & Morrill Inc., New York, NY
Client: Nestle Foods Corporation, White Plains, NY

Endorsed under the Nestea umbrella name, this new, flavored drink mix has a package design that emphasizes strong brand identification, a color-coding system for each flavor, and appetite-appealing photography.

Product: Moul—bie Flour
Designer: Alinea Graphic
Design Firm: Alinea Graphic, Paris, France
Client: Grands Moulins de Paris, France

To re-establish this product under its new brand name, Grands Moulins de Paris redesigned the package and informed the consumer that the flours have been prepared for specific baking needs.

Product: Voiello Pasta
Designer: Gió Rossi
Design Firm: Image Plan International, Milano, Italy
Client: Voiello, Italy

Voiello semolina pasta is a leading brand in the high-priced segment of the market. This brand is sold exclusively in Italy and is packaged in cellophane. The logo represents Pulcinella, the carnevalesque mask of Naples, and the bay of the city.

Product:	Chef Tell's®Pasta PourOvers™
Creative Director:	Barry Seelig
Design Manager:	Lou Luscher
Designers:	Richard Karsten, Jim Best, Jane Hord, and Monica Kurkemelis
Design Firm:	Apple Design Source Inc.
Client:	Madisons Inc.

The dominant black color on the label gives the product high shelf impact because of its distinctiveness and conveys a top-quality image. Chef Tell's face appeals to many consumers in the target audience who recognize him, and it helps identify the product with the person who developed it. The bottle shape conveys quality and uniqueness, and makes the bottle easier to handle.

Product: Gastone Saclá Recipes
Designer: Gió Rossi
Design Firm: Image Plan International, Milano, Italy
Client: Saclá, Italy

Home-like quality is communicated by quaint country styling, typography, and illustration.

Product: Crabtree & Evelyn Crackers
Designer: Peter Windett & Associates
Design Firm: Peter Windett & Associates, London, England
Client: Crabtree & Evelyn, London, England

Crabtree & Evelyn Italian wheat crackers are packaged in boxes with old-fashioned illustrations, and drawings of wheat to signify the ingredients of the crackers.

Product: Sinosmina Breath Fresheners
Designer: Kelemata
Design Firm: Kelemata, Torino, Italy
Client: Kelemata, Torino, Italy

The main feature of this breath freshener package is the informative copy, which explains the different properties of herbs and their ability to freshen the mouth.

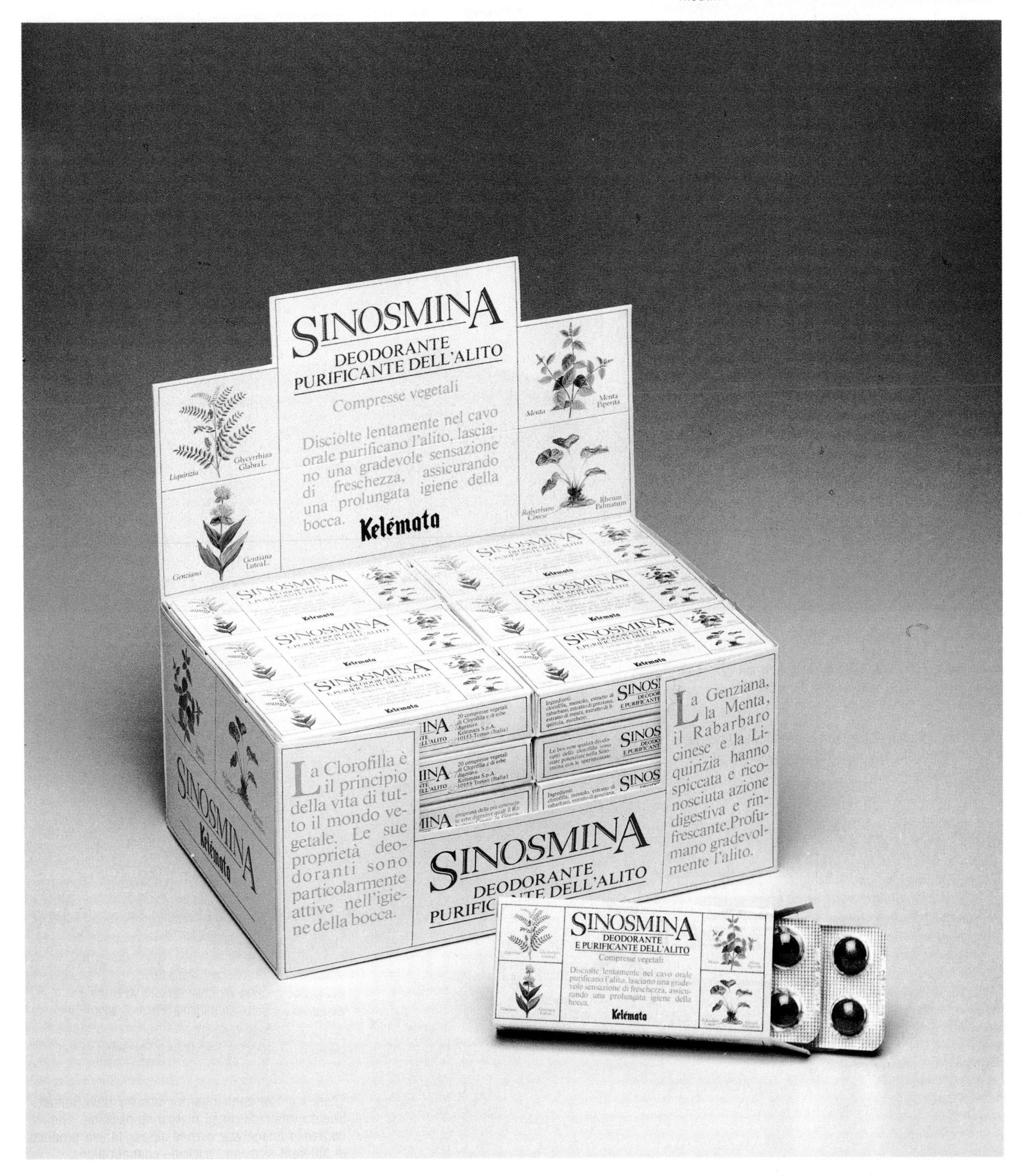

Product: Dieta Sugar Substitute
Designer: Lisa Caputo
Design Firm: The Creative Source Inc., New York, NY
Client: Rochala Corporation, Brooklyn, NY

The bi-lingual copy for this new product appeals to a Latin/American, younger, health conscious and athletic market.

Product: Blue Bonnet Margarine
Designers: Owen W. Coleman and John Rutig
Design Firm: Coleman, LiPuma, Segal & Morrill Inc., New York, NY
Client: Nabisco Brands Inc., East Hanover, NJ

There's a new contemporary look for Blue Bonnet Sue on this redesigned margarine package. The upgraded image was carried across fifteen products in different sizes and package configurations.

Product: Celentano Frozen Entrees
Designers: Judith Miller and Sabra Kushlefsky
Design Firm: Gerstman & Meyers Inc., New York, NY
Client: Celentano Bros., Verona, NJ

Celentano's red logo on a white background creates premium imagery that conveys a contemporary appetite appeal.

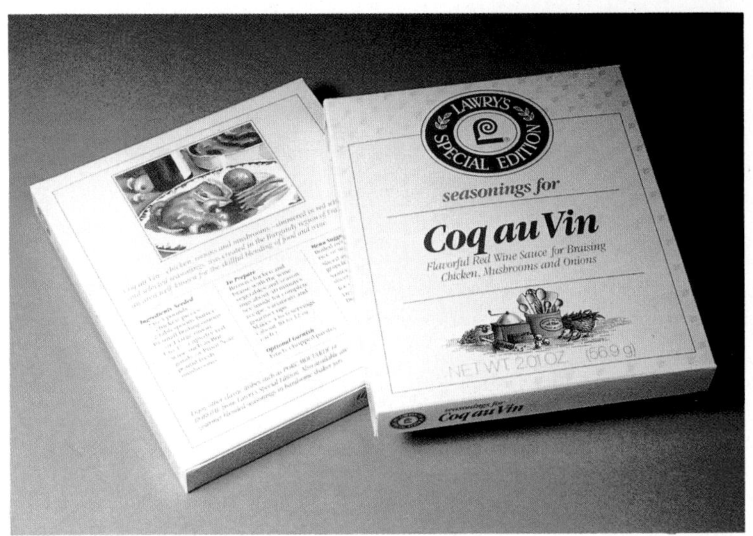

Product: Lawry's Special Edition Sauces
Designers: Art Goodman and Chuk Yee-Cheng
Design Firm: Bass/Yager & Associates, Los Angeles, CA
Client: Lawry's

Classic gourmet styling emphasizes Lawry's Special Edition Sauces. The bold typography labels each individual Sauce package by flavor.

Product: Chicken Kiev
Designer: Mary Lewis
Design Firm: Lewis Moberly, London, England
Client: Buxted Poultry Ltd., Yorkshire, England

This package was designed to create impact in the freezer case with colorful graphics.

Product: Houston's Jarred and Canned Nuts
Designer: Karen Corell
Design Firm: Gerstman & Meyers Inc., New York, NY
Client: Peanut Processors Inc., Dublin, NC

The Houston's line of jarred and canned nuts captures its desired adult audience by conveying a festive environment. Color-coded nut varieties and strong product distinction separates it from competitors.

Product: Tang
Designer: Peter Nikolits
Design Firm: General Foods Corporation—Corporate Design Center, White Plains, NY
Client: General Foods Corporation, White Plains, NY

Energetic illustration sparks this new image for Tang.

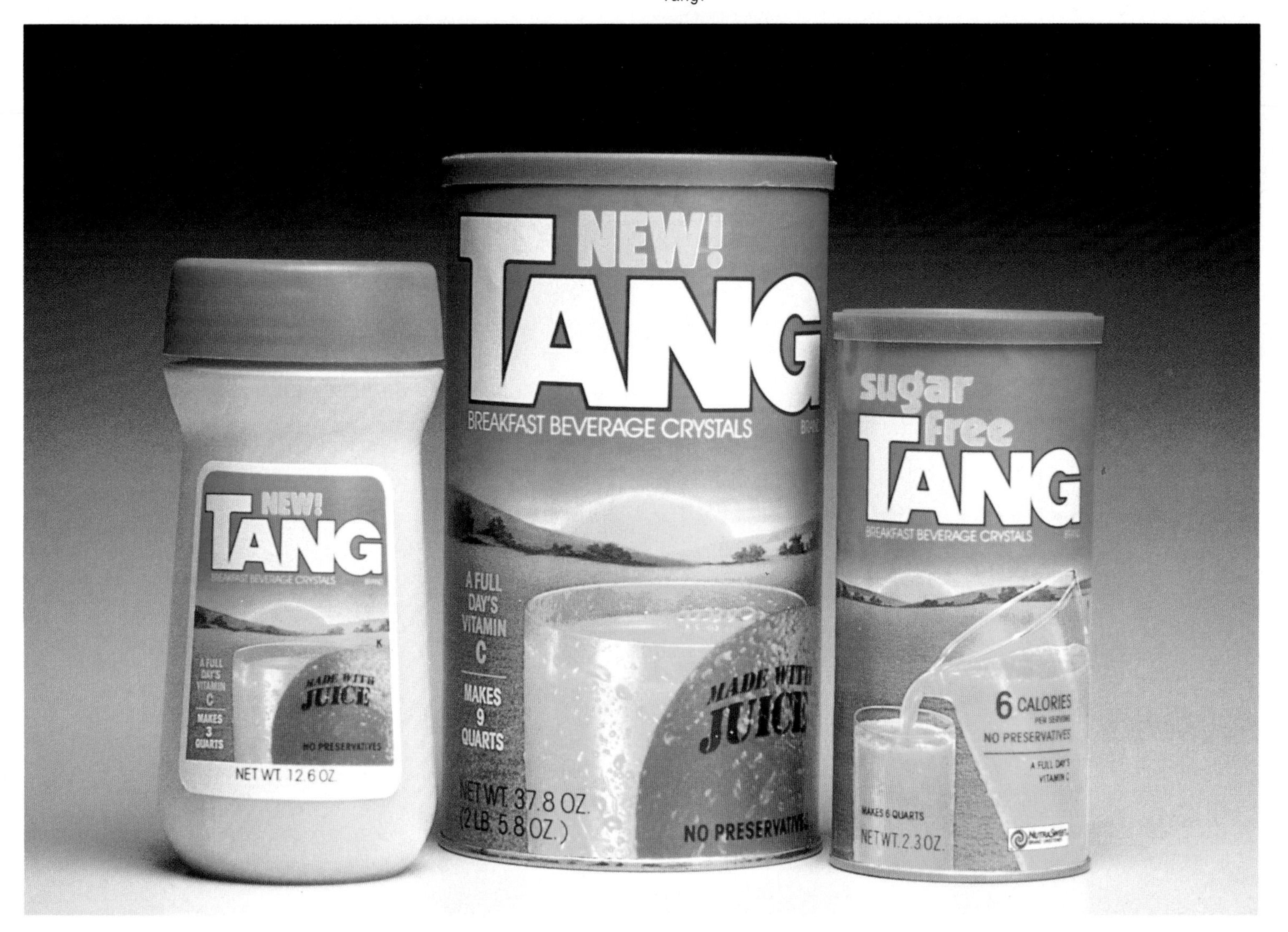

Product: French's Colemans Wet Mustard
Designer: Dixon & Parcels Associates, Inc.
Design Firm: Dixon & Parcels Associates, Inc., New York, NY
Client: RT French Company, Rochester, NY

The redesign of the Coleman Wet Mustard line uses, with special permission, the English Royal Warrant and Coat of Arms. A handsome gold collar adds to the perception of quality and enhances the unified appearance of the overall line. Easy identification is made possible by color-coded ovals.

Product: Lipton Long Grain Wild Rice
Designer: Dixon & Parcels Associates, Inc.
Design Firm: Dixon & Parcels Associates, Inc., New York, NY
Client: Thomas J. Lipton Inc., Englewood Cliffs, NJ

This line of Lipton's Rice & Sauce features Long Grain & Wild Rice with four seasonings. The strong typography is enhanced by a dominant photograph of a serving suggestion.

Product: Slim Jim
Designer: Dixon & Parcels Associates, Inc.
Design Firm: Dixon & Parcels Associates, Inc., New York, NY
Client: Goodmark Foods Inc., Raleigh, NC

The redesign of this line of old-fashioned beef sticks focuses on appealing to the sportsman. Sketches of various sports activities, combined with a color slash, assist consumers in choosing their favorite flavors.

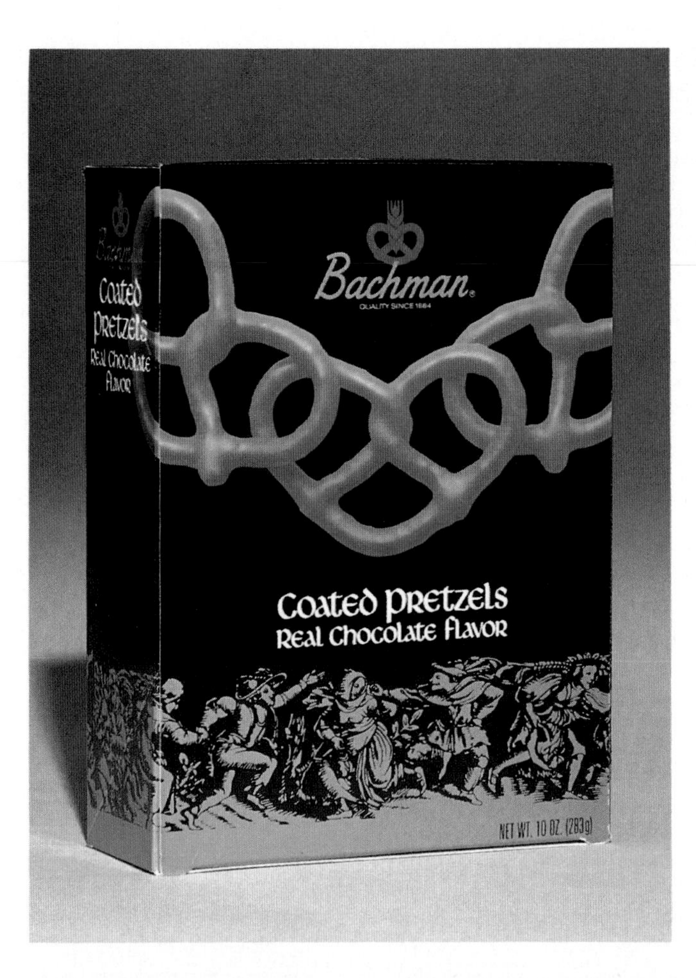

Product: Bachman Chocolate-Covered Pretzels
Designer: Dixon & Parcels Associates, Inc.
Design Firm: Dixon & Parcels Associates, Inc., New York, NY
Client: The Bachman Company, Reading, PA

This packaged version of a Pennsylvania Dutch favorite uses rich photography to whet the appetite of the consumer. The dark background creates an appealing and dramatic shelf presence.

Product: Bachman Hard Pretzels
Designer: Dixon & Parcels Associates, Inc.
Design Firm: Dixon & Parcels Associates, Inc., New York, NY
Client: The Bachman Company, Reading, PA

Full-color, appetizing photography brings mouth-watering attention to Bachman Hard Pretzels. Strong typography against the dark background projects a dynamic image, enhancing the total impression.

Product: French's Worcestershire Sauce
Designer: Dixon & Parcels Associates, Inc.
Design Firm: Dixon & Parcels Associates, Inc., New York, NY
Client: RT French Company, Rochester, NY

To raise the perceived value of French's Worcestershire Sauce, the RT French Company chose a label enhanced with photographs of appetizing food. To complete the package, a dark neckband was created to give an appearance of richness and flavor.

Product: Mulino Bianco
Designer: Gió Rossi
Design Firm: Image Plan International, S.p.A., Milan, Italy
Client: Barilla G. & B. F.lli S.p.A., Italy

Mulino Bianco, a brand that today includes seven product lines, was introduced ten years ago on the Italian market. Starting with a line of cookies, the brand now has branched into breads, snacks and industrial pastry, and it is a market leader in all categories. Mulino Bianco means "white mill," indicating a fresh, wholesome product made exclusively from natural ingredients. Mulino Bianco was the first brand to feature reclosable bags, a feature often imitated by competing brands.

Product: Christmas Packages
Designer: Amy Leppert
Design Firm: Murrie White Drummond & Lienhart Associates, Chicago, IL
Client: McDonald's Corporation

This seasonal package communicates the unique proprietary "feel" of quality and warmth that is associated with "the McDonald's experience."

Product: McDonald's Cookies
Designer: Amy Leppert
Design Firm: Murrie White Drummond & Lienhart Associates, Chicago, IL
Client: McDonald's Corporation

The McDonaldland cartoon characters take on a slightly more stylized form here. The McDonald's image of quality and fun is represented by the graphics.

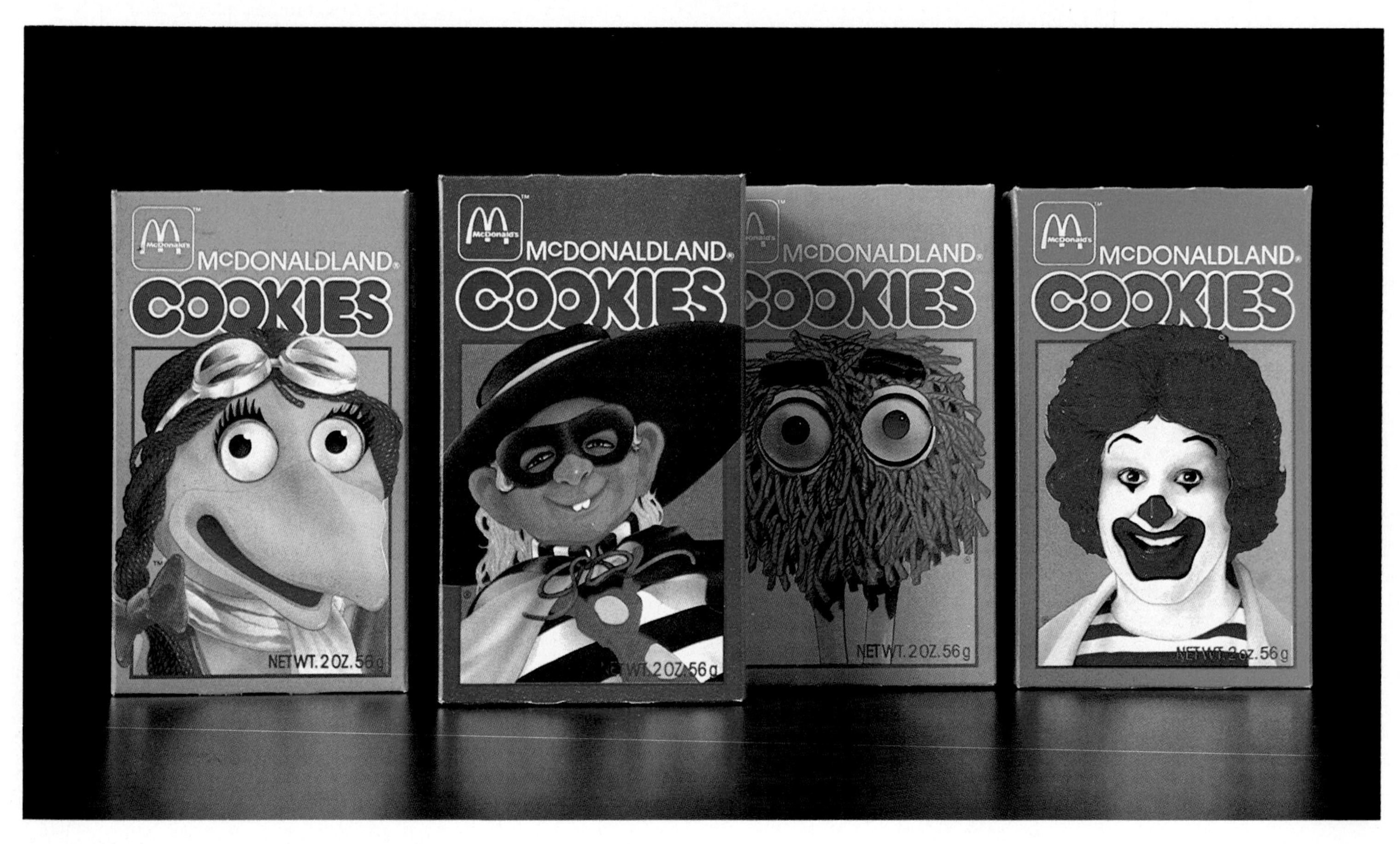

Product: Harbor Sweets
Designer: Dixon & Parcels Associates, Inc.
Design Firm: Dixon & Parcels Associates, Inc., New York, NY
Client: Harbor Sweets Inc., Salem, MA

To motivate the catalog mail order buyer, this design was created using simple, rich red complemented by gold. It projects the elegance and richness of Harbor Sweets candies.

Product: Mrs. Paul's Au Naturel Fish Fillets
Designer: Dixon & Parcels Associates, Inc.
Design Firm: Dixon & Parcels Associates, Inc., New York, NY
Client: Mrs. Paul's, a subsidiary of Campbell Soup Company, Philadelphia, PA

Full-color photography was used to suggest ways of serving each type of Mrs. Paul's 100% natural fish fillets. Positioned against a rich background, the fillets project a truly realistic image designed to motivate the consumer.

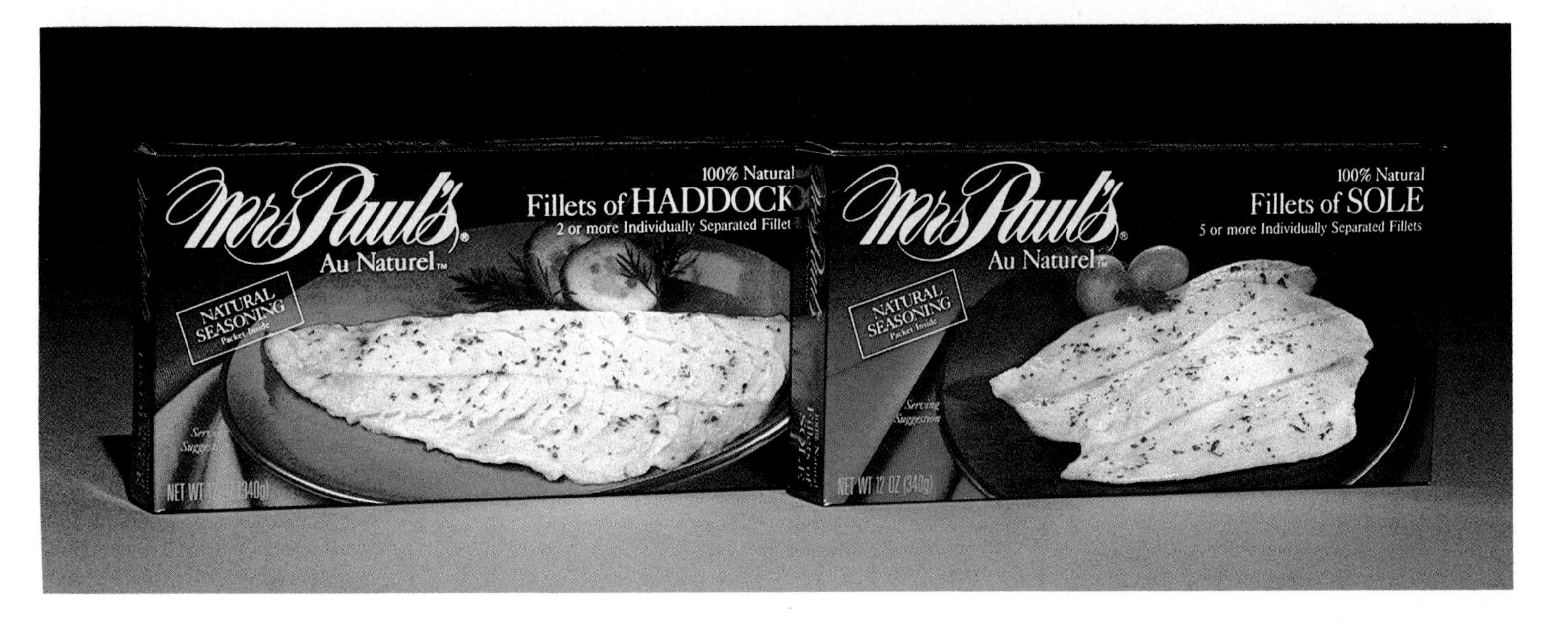

Product: Mont Blanc Crème Dessert
Designer: Shining
Design Firm: Shining, France
Client: Gervais Danone, France

All Mont Blanc flavors are packaged in the same can, but each flavor is assigned its own color. Illustrations on each label help identify the flavors.

Product: Uncle Ben's Long Grain and Wild Rice
Designer: Kornick Lindsay
Design Firm: Kornick Lindsay, Chicago, IL
Client: Uncle Ben's Inc.

The packaging of Uncle Ben's rices is not only color-coded, but each package features a photograph of the kind of rice it contains. When linked together visually, the packages create a strong billboard effect.

Product: French's Pasta Toss Seasoned Cheese Topping
Designer: Kornick Lindsay
Design Firm: Kornick Lindsay, Chicago, IL
Client: RT French Company, Rochester, NY

The graphics emphasize the light nature of this pasta topping. Foil label stock is used to brighten the color-coded flavor system.

Product: Burger King/Chicken Tenders
Designers: Alex Pennington and Harriet Pertchik
Design Firm: Gerstman & Meyers Inc., New York, NY
Client: Burger King Corporation, Miami, FL

This new Chicken Tenders package is designed to look like a picnic basket wrapped in a vermilion grosgrain ribbon, with an oval emblem on the top panel. The inner oval of the Chicken Tenders logo bears a symbol of a plump chicken.

Product:	Naturell Yoghurt
Designer:	Tom Hedqvist
Design Firm:	Hall & Cederquist Ad Agency, Sweden
Client:	The Swedish Dairy Cooperative
Awards:	1986 Outstanding Design Award, 1986 Outstanding Swedish Form Award

This is certainly a different approach to yogurt packaging. The intention here is to associate yogurt with health and long life.

Product: Mulino Bianco Cookies
Designer: Gió Rossi
Design Firm: Image Plan International, Italy
Client: Barilla G. & B. F.lli S.p.A., Italy

These high-quality, reasonably priced cookies come in a festive, easy-to-use package.

Product: Kelloggs' Nutri-Grain
Designer: Mary Lewis
Design Firm: Lewis Moberly, London, England
Client: Kelloggs', Manchester, England

Kelloggs' largest-ever product launch breaks with traditional corporate styling to give more sensitive and individual branding to Nutri-Grain, a cereal intended for adults with a taste for the natural.

Product: Shake 'N Bake
Designer: Raymond Woolard
Design Firm: General Foods Corporation—
Corporate Design Center, White Plains, NY
Client: General Foods Corporation, White Plains, NY

The Shake 'N Bake relaunch features a bright, colorful look for each product in the line. Colorful, color-coded boxes are easily distinguished.

Product: Red Cheek Juices
Designers: Scott Johnson, Regina Rubino, and Lisa Lien
Design Firm: Gerstman & Meyers Inc., New York, NY
Client: Red Cheek Division, Fleetwood, PA

New label graphics and new bottle shape were developed for Red Cheek juices. The packaging features high-quality photography of a big, fresh, dewy apple, a new logo, and color-coded backgrounds. The new appearance enhances Red Cheek's juice products.

Product: La Glace Ice Cream
Designer: Shining
Design Firm: Shining, France
Client: Gervais Danone, France

These ice cream and sherbert containers are much different than ordinary containers. They enable the ice cream or sherbert to be stored freshly after opening. Appetizing photography and the bright red Gervais logo is a striking factor to this product's success.

Product: Post Fruit & Fibre Cereal
Designers: Sal LiPuma and Abe Segal
Design Firm: Coleman, LiPuma, Segal & Morrill Inc., New York, NY
Client: General Foods Corporation, White Plains, NY

A new design image reflecting a strong Fruit & Fibre story was carried across four products, together with color-coded backgrounds in strong, bright colors.

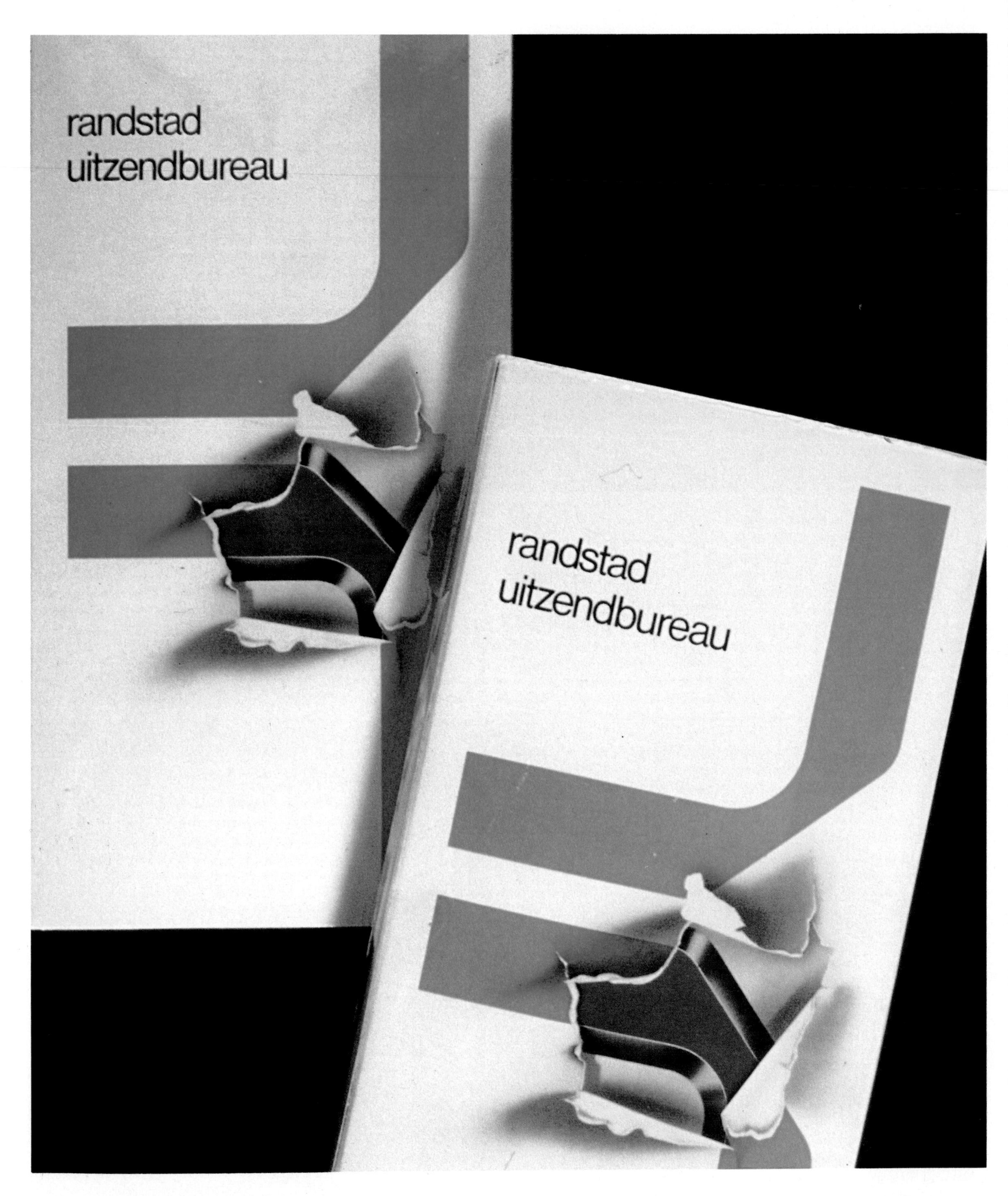

Product: Box of Chocolates
Designer: Frans Lieshout
Design Firm: Total Design, Amsterdam
Client: Randstad Uitzendbureau bv

This new, clever design method was used for a box of chocolates.

Product: Good 'n Buttery Biscuits
Designer: Kornick Lindsay
Design Firm: Kornick Lindsay, Chicago, IL
Client: Pillsbury

The use of foil on this package gives impact to the new product and distinguishes Pillsbury Biscuits from others in its market.

Product: Baker's Chocolate and Coconut
Designer: Raymond Woolard
Design Firm: General Foods Corporation—Corporate Design Center, White Plains, NY
Client: General Foods Corporation, White Plains, NY

Baker's employs a wide use of colors, package sizes, and package types to achieve a unified image.

Product: Natural Yogurt
Designer: Design Group Italia
Design Firm: Design Group Italia, Milano, Italy
Client: Ala Zignago, Milano, Italy

More attractive than plastic jars, these glass containers allow for a highly visible product. The bottle shape is designed so that spooning out the product is easy.

Product: Il Fornaio Nutritious Line
Designer: Design Group Italia
Design Firm: Design Group Italia, Milano, Italy
Client: Servicepan, Italy

The Nutritious Line consists of assorted breads, snacks, spaghetti, and rice. Each package is designed to identify its particular product. All packages are sealed for freshness and are easy to reseal after opening.

Product: Henri Nestle Chocolate
Designer: Lister Butler Inc.
Design Firm: Lister Butler Inc., New York, NY
Client: Nestle

This eye-catching design for Henri Nestle provides a distinctive image for lovers of fine chocolate. Color-coding helps distinguish chocolate types.

Product: Bee House Spice Canister
Designer: Takeo Ino
Design Firm: Bee House Commercial Goods Research Institute, Tokyo, Japan
Client: Bee House Company Ltd., San Francisco, CA

This Bee House spice canister gift package is a versatile, handy kitchen item.

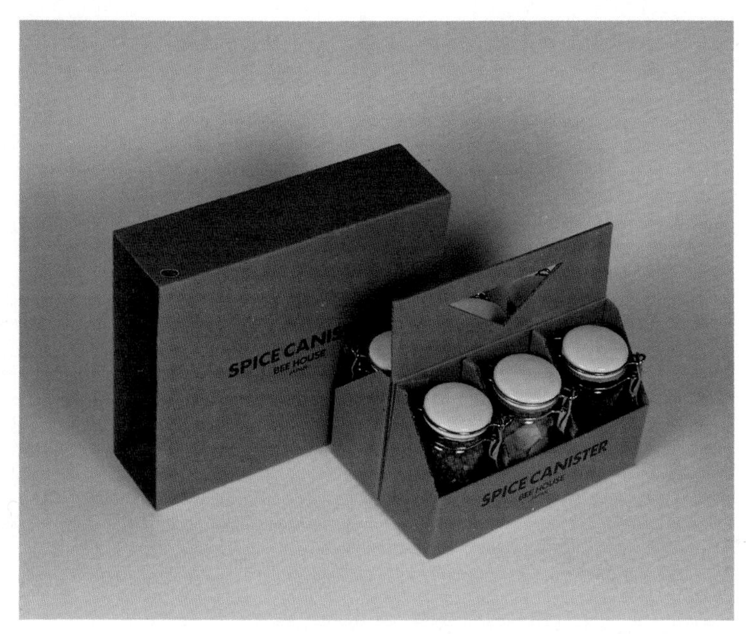

Product: Barilla Pasta
Designer: Michael Peters
Design Firm: Michael Peters and Partners, London, England
Client: Barilla G. & B. F.lli S.p.A., Italy

When Barilla packaged pasta in cardboard boxes in the 1950s, it upgraded the brand image. The change was originally made because the brand image is more visible on boxes than on cellophane. Deviating from other brands packaged in cellophane, Barilla took the chance of packaging differently and was highly successful. Today Barilla owns twenty-five percent of the market.

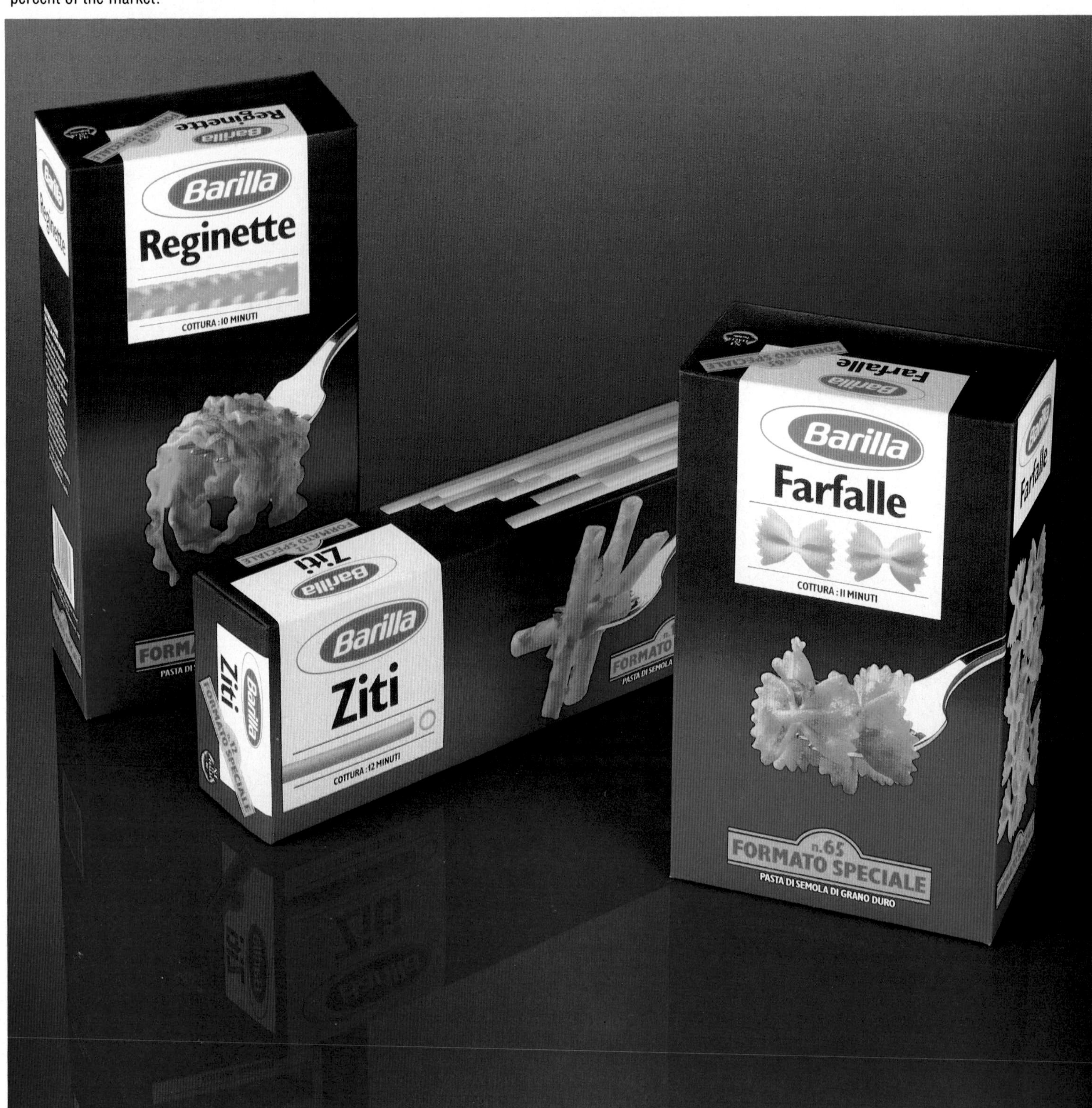

Product: Orville Redenbacher Gourmet Popping Corn
Designers: Owen W. Coleman, John Rutig, and Abe Segal
Design Firm: Coleman, LiPuma, Segal & Morrill Inc., New York, NY
Client: Beatrice/Hunt-Wesson Inc., Fullerton, CA

The Orville Redenbacher image was updated by combining a bold stylish brand name logo printed on a subtle farm scene background, with an upbeat illustration of Orville.

Product: Aunt Jemima Waffles
Designers: Owen W. Coleman, Abe Segal, Ward Hooper, and Cathy Sze-Tu
Design Firm: Coleman, LiPuma, Segal & Morrill Inc., New York, NY
Client: The Quaker Oats Company, Chicago, IL

This redesign program for all Aunt Jemima Frozen Waffles was executed with computer graphics at every creative phase of development. Quaker was seeking a new, fresh image that would better set its products apart from other brands. Criteria included a more uniform line design and a continued highlighting of the Aunt Jemima syrup bottle.

Product: Sperlari Candy
Designer: Sandro Gramolelli
Design Firm: Armando Testa S.p.A., Italy
Client: Sperlari F.lli S.p.A., Italy

Sperlari uses a cute, new package to add a bit of fun to their candy.

Product: Musici Chocolates
Designer: Giampiero Ferrari
Design Firm: Armando Testa S.p.A., Italy
Client: Sperlari F.lli S.p.A., Italy

This box of fine chocolates is meant for special occasions. Written in gold characters are the phrases "Offer them to the sound of music" and "Open and listen to the soft notes of Silent Night."

Product: L'Arte Pasticcera Bread
Designer: Design Group Italia
Design Firm: Design Group Italia, Milano, Italy
Client: Ruggero Bauli S.p.A., Italy

Bauli makes different varieties of panettone. This package was designed for their top-of-the-line bread.

Product: Durkee Olives
Designers: Sal LiPuma and John Rutig
Design Firm: Coleman, LiPuma, Segal & Morrill Inc., New York, NY
Client: Durkee, Westlake, OH

Durkee redesigned their entire line of olive products with this bright, new, fresh image.

Product: Top That Popcorn Spray
Designers: Owen W. Coleman and Abe Segal
Design Firm: Coleman, LiPuma, Segal & Morrill Inc., New York, NY
Client: The Dial Corporation, Phoenix, AZ

The Dial Corporation produced a new product projection, a fun image for this aerosol popcorn spray available in two flavors.

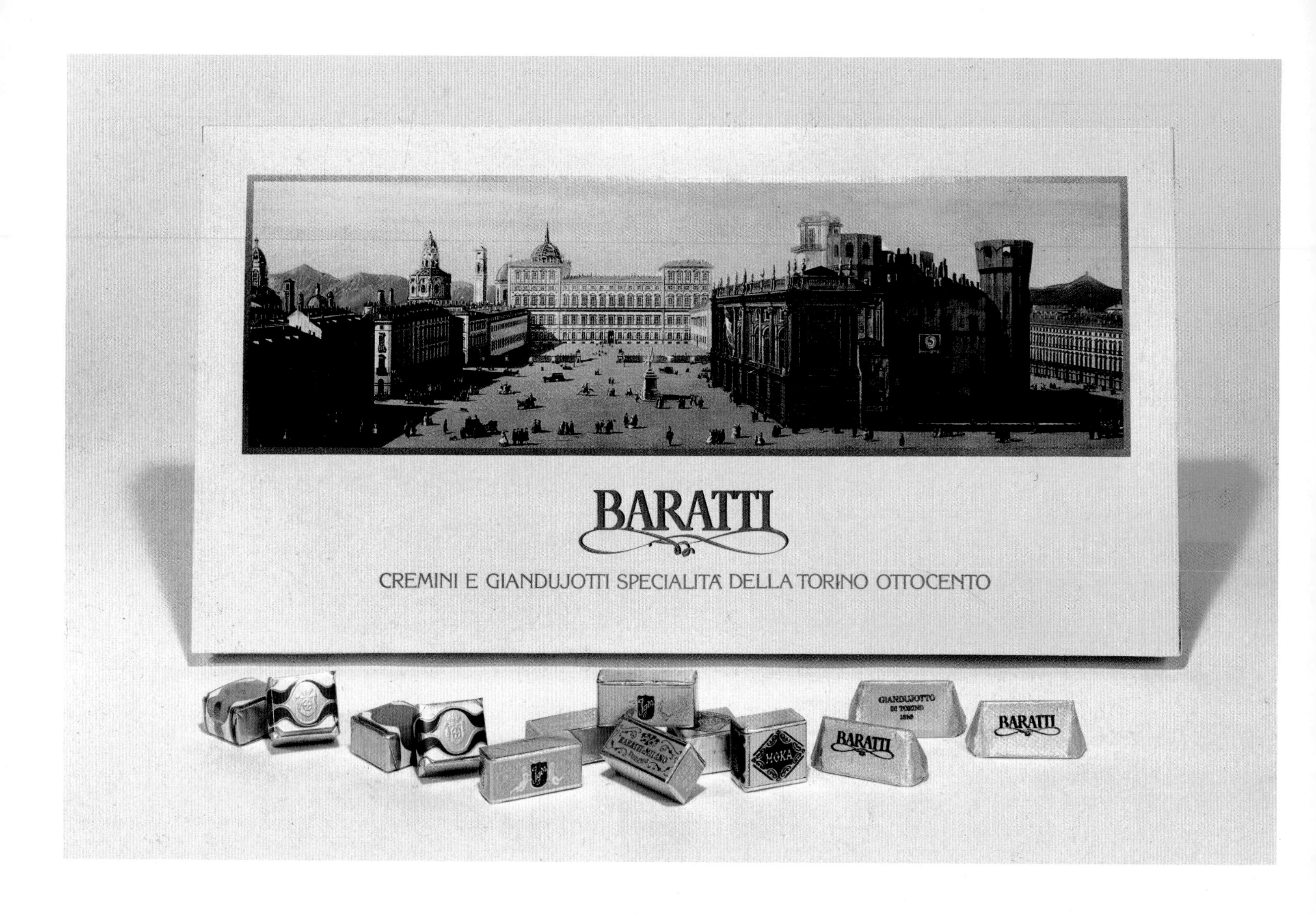

Product: Baratti & Milano Candies and Chocolates
Client: Baratti & Milano, Torino, Italy

The traditional design of Baratti chocolate packages reflects the history of the century-old company. Boxes for assorted chocolates feature a view of Piazza Castello in Torino, and the box of Giandujotti, named after a carnival figure, is illustrated with his portrait.

CARAMELLE
BARATTI
MENTA
ARANCIA
FRAGOLA
NOCCIOLA
RATAFIA
BARATTI
LE CARAMELLE di TORINO
MINT DROP
RIPIENA
mango

Product: Breyers Ice Cream
Designer: Larry Riddell
Design Firm: Gertsman & Meyers Inc., New York, NY
Client: Kraft Inc. Dairy Group, Philadelphia, PA

The photography as well as the typography of these new Breyer packages creates an assurance of a natural, healthy product, backed by an established brand name logo.

Product: Salerno Cookies
Designer: Ward Hooper
Design Firm: Coleman, LiPuma, Segal & Morrill Inc., New York, NY
Client: General Biscuit Brands, Elizabeth, NJ

This package design for a line of Salerno Cookies uses a strong typographical format and is produced on clear cello wrap.

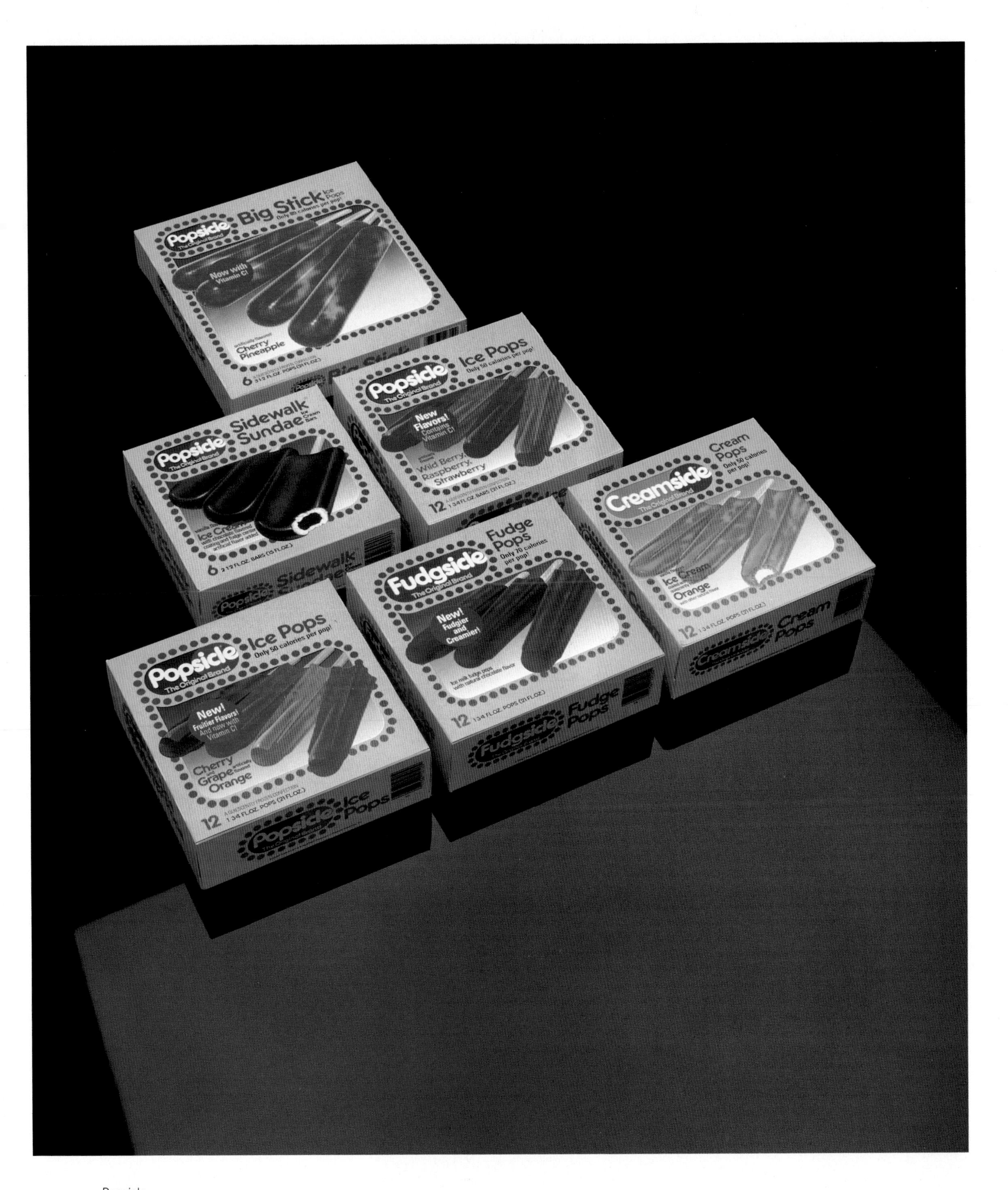

Product: Popsicle
Designer: Popsicle Inc.
Design Firm: Popsicle Inc., Englewood, NJ
Client: Popsicle Inc., Englewood, NJ

Bright, appetizing colors set off each popular flavor in this new design for Popsicle.

Product: Cobblestone Mill
Designer: Dixon & Parcels Associates, Inc.
Design Firm: Dixon & Parcels Associates, Inc., New York, NY
Client: Flowers Industries Inc., Thomasville, GA

Establishing this line of six different thinly sliced breads required the creation of a new name and a graphic presentation of that name. The Cobblestone Mill conveys the rich quality of the breads while color-coding and strong typography differentiate the types of bread for the consumer.

Product: Mrs. Paul's Seafood Encroute
Designer: Dixon & Parcels Associates, Inc.
Design Firm: Dixon & Parcels Associates, Inc., New York, NY
Client: Mrs. Paul's, a subsidiary of Campbells Soup Company, Philadelphia, PA

To project the flavorful nature of these French classic seafood entrees, the design features a photograph of a fork breaking the pastry. A light background enhances the product's light nature, and the Mrs. Paul's "script" logo is strategically placed across the end of the package.

Product: Brown Gold Coffee
Designers: Owen W. Coleman and Abe Segal
Design Firm: Coleman, LiPuma, Segal & Morrill Inc., New York, NY
Client: Tetley Inc., Shelton, CT

A new image was created for all Brown Gold coffee products, both regular and decaffeinated.

Product: Coffee In A Filter
Designer: Dixon & Parcels Associates, Inc.
Design Firm: Dixon & Parcels Associates, Inc., New York, NY
Client: C.B. Rafetto, Stamford, CT

Coffee In A Filter packages are designed with an over-all coffee bean background. Color bands help identify the coffee type.

Product: Maxwell House Private Collection
Designer: Peter Nikolits
Design Firm: General Foods Corporation—Corporate Design Center, White Plains, NY
Client: General Foods Corporation, White Plains, NY

A fine script logo creates new appeal from the coffee industry, as seen in the Maxwell House line of Private Collection coffees.

Product: Thomas Garroway
Designer: Raymond Sixsmith
Design Firm: General Foods Corporation—Corporate Design Center, White Plains, NY
Client: General Foods Corporation, White Plains, NY

These General Foods packages achieve classic product representation by using traditional styling.

Product: Lipton's Rice and Sauce
Designer: Dixon & Parcels Associates, Inc.
Design Firm: Dixon & Parcels Associates, Inc., New York, NY
Client: Thomas J. Lipton Inc., Englewood Cliffs, NJ

These five seasoned rice and sauce dishes are packaged in foil pouches for freshness and for easy display. Full-color photographs help achieve a strong retail impact and appetite appeal.

Product: Carlton Cheese
Designers: Peter Hemann and Norbert Hrdliczka
Design Firm: Design Board/Behaeghel & Partners S.A., Belgium
Client: Gervais Danone AG, Germany

This new design concept for a white cheese expresses exclusiveness and quality.

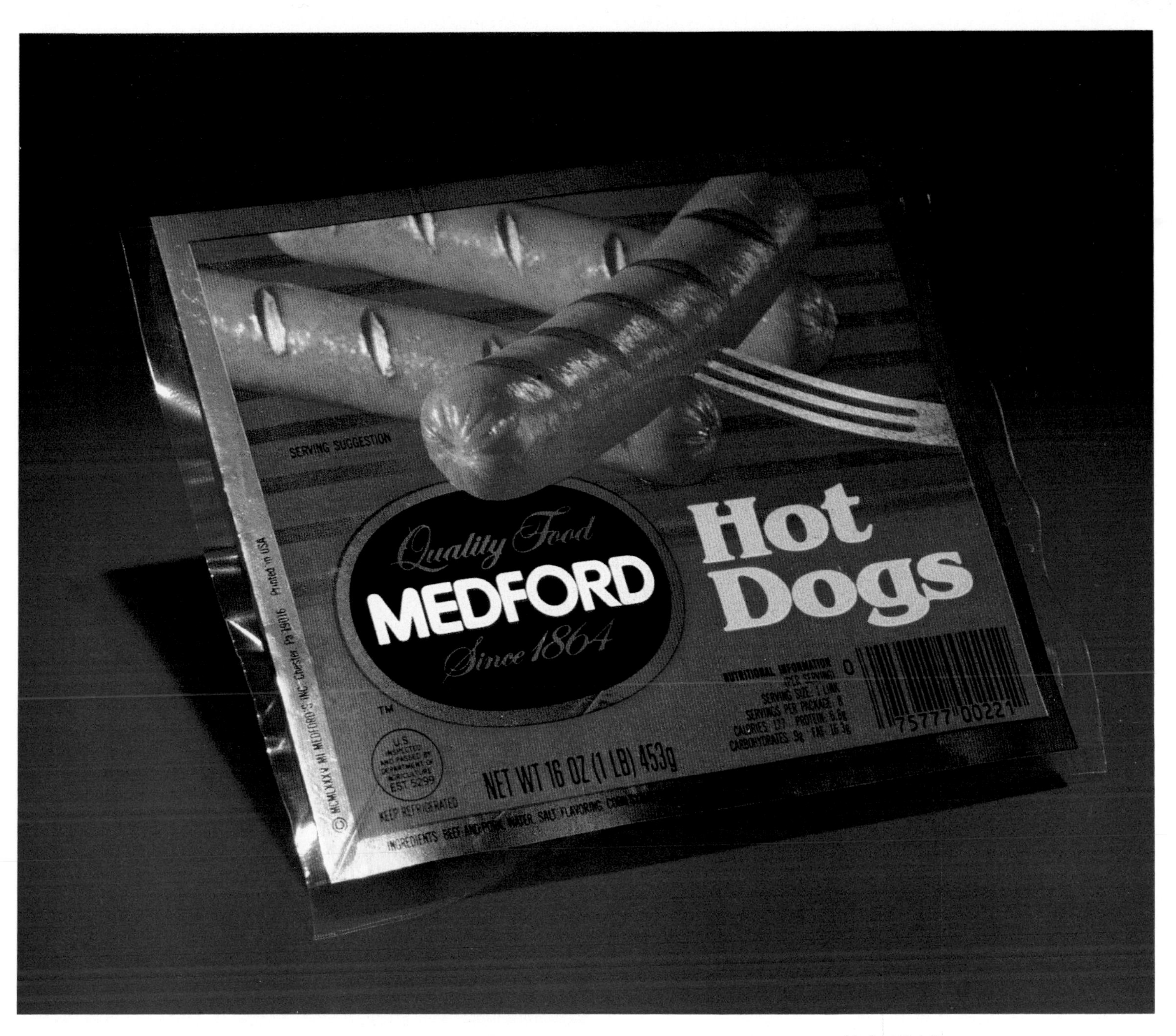

Product: Medford Hot Dogs
Designer: Dixon & Parcels Associates, Inc.
Design Firm: Dixon & Parcels Associates, Inc., New York, NY
Client: Medford Inc., Chester, PA

For appetite appeal, this Medford Hot Dog package shows a full-color photograph of freshly grilled hot dogs. The rich background colors enhance the entire presentation while strong typography establishes the brand name for quick identification in the freezer case.

Product: Entenmann's Lite
Designer: Norman Cohen
Design Firm: General Foods Corporation—Corporate Design Center, White Plains, NY
Client: General Foods Corporation, White Plains, NY

A crisp new look for Entenmann's Lite Products, which uses a brisk blue on a white background, appeals to a specific target audience.

Product: Pillsbury/Azteca Mexican Foods
Designers: Alex Pennington
Design Firm: Gerstman & Meyers Inc., New York, NY
Client: The Pillsbury Company, Minneapolis, MN

The golden sun symbol and bold logotype on dark terracotta backgrounds help Pillsbury achieve an air of festive ambiance for their new line of fresh, premium refrigerated Mexican food products.

Product:	Conzello Salad Dressings
Designer:	Kornick Lindsay
Design Firm:	Kornick Lindsay, Chicago, IL
Client:	Kraft Inc.

A new bottle shape and graphics were designed for a new line of authentic Italian dressings. The glass is embossed with a special seal of quality and the graphics reflect the bold color and simple design that distinguishes authentic Italian packaging.

Product:	Lady Aster Frozen Foods
Designers:	John Neher and Edward Rebek
Design Firm:	John Racila Associates, Oak Brook, IL
Client:	Culinary Foods Inc., Chicago, IL

Dramatic and appetizing photography was combined with a distinctive silver carton to convey the gourmet quality and attractiveness of the product.

Product: Deans Farm Eggs
Designers: Philip Griffiths and Tony Hatt
Design Firm: Leslie Millard Associates, London, England
Client: Deans Farm Eggs Limited, England

Freshness is indicated by use of delicate watercolors on these egg cartons.

Product: Post Horizon Cereal
Designers: Sal LiPuma, Abe Segal, and John Rutig
Design Firm: Coleman, LiPuma, Segal & Morrill Inc., New York, NY
Client: General Foods Corporation, White Plains, NY

The Post Horizon package communicates to young, affluent, health-conscious consumers that the cereal's good taste is accomplished by using only natural ingredients. General foods achieves this with its strong graphic image.

Product: Fazer Wander Egg
Designers: Lewis Lowe and Diane Nelson
Design Firm: S & O Consultants Inc., San Francisco, CA
Client: Fazer Chocolates Inc., Los Angeles, CA

This Fazer Chocolates package has a clever design and makes a delightful gift as well as an eye-catching shelf display.

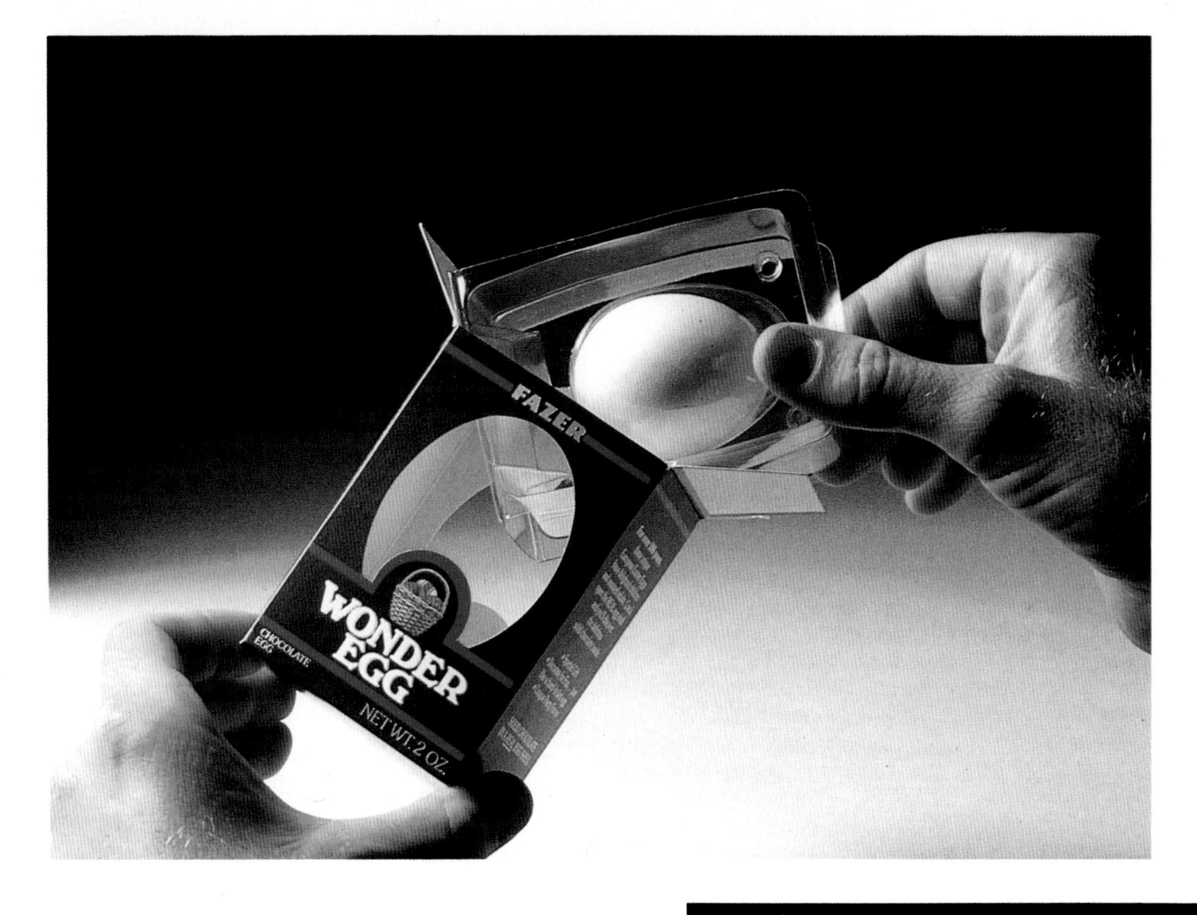

Product: LU Cookies, Chips Chocolat, and Aloha
Designers: Owen W. Coleman and Ward M. Hooper
Design Firm: Coleman, LiPuma, Segal & Morrill Inc., New York, NY
Client: General Biscuit Brands Inc., Elizabeth, NJ

When General Biscuit Brands established a new packaging design program for over twenty LU brand products, it also introduced two new product flavors. The Chips Chocolat has been an outstanding new product success.

Product: Babicao Instant Breakfast
Designer: Shining
Design Firm: Shining, France
Client: Nestle, France

Nestle's children's drink, a low-fat blend of cocoa and cereal, is packaged in a box that appeals to young children. Shown on the label is an illustration of corn and cocoa beans.

Product: Chunky
Designers: Owen W. Coleman and Abe Segal
Design Firm: Coleman, LiPuma, Segal & Morrill Inc., New York, NY
Client: Nestle Foods Corporation, White Plains, NY

Nestle Foods reintroduced its Chunky candy bar in the trapazoid configuration. The new, upgraded design image utilizes a bold logo positioned on a silver striped background.

Product: Crosse & Blackwell Soups
Designer: Owen W. Coleman
Design Firm: Coleman, LiPuma, Segal & Morrill Inc., New York, NY
Client: Nestles Food Corporation, White Plains, NY

Appetizing photography was executed for eighteen Crosse & Blackwell soups. Representative color-coding and coordinating backgrounds were also used for each hearty soup.

Product: Wishbone Lite Salad Dressing
Designers: Owen W. Coleman and Abe Segal
Design Firm: Coleman, LiPuma, Segal & Morrill Inc., New York, NY
Client: Thomas J. Lipton Inc., Englewood Cliffs, NJ

A major design change across all Lite Wishbone dressing labels reflects an upscale contemporary image. The unique bulls-eye design utilizes color-coding for each flavor.

Product: Farmer's Choice Butter Pack
Designer: Suzanne Perlston
Design Firm: Leslie Millard Associates, London, England
Client: Eilers & Wheeler, London, England

A crisp gold-foil wrap and a bold blue logo communicate freshness and brand identity.

Product: Wispride Cheese Spread
Designers: Owen W. Coleman, John Rutig, and George Beckstead
Design Firm: Coleman, LiPuma, Segal & Morrill Inc., New York, NY
Client: Nestle Foods Corporation, White Plains, NY

This design, developed to create a new image, was carried across five cheese flavors. A cardboard sleeve structure was developed to house the individual tubs containing the product.

Celeste®
PIZZA·FOR·ONE™
Calcium propionate added to retard spoilage of crust.
New Improved Crust!
Serving Suggestion
KEEP FROZEN
NET WT 7½ OZ • 212g
Vegetable WITH MUSHROOMS, GREEN & RED PEPPERS, ONIONS & OLIVES...NO MEAT!

Now Mama Celeste's Famous Pizza is Better Than Ever!
You can still enjoy the great pizza taste that Mama Celeste made famous back in the 1930's.
Celeste has everything a great pizza should have: lots of cheese and luscious topping, zesty sauce, Italian seasonings...and now a new light and crispy crust.
Celeste PIZZA
MICROWAVEABLE
DIRECTIONS:
FOR A CRISP CRUST:
1. Place cookie sheet on center oven rack.
2. Preheat oven to 400°F.
3. Remove pizza from carton; remove plastic wrap.
4. Place frozen pizza on PREHEATED cookie sheet.
5. Bake 13 to 15 minutes or until center cheese is melted and crust edge is golden brown.
FOR AN EXTRA CRISP CRUST:
1. Preheat oven to 400°F.
2. Remove pizza from carton; remove plastic wrap.
3. Place frozen pizza directly on center oven rack.
4. Bake 11 to 13 minutes or until center cheese is melted and crust edge is golden brown.
NOTE: Due to variations among ovens, baking times may differ. To help keep oven clean, place aluminum foil on oven bottom before heating. Check oven manufacturer's instructions for use of foil.
MICROWAVE DIRECTIONS*:
1. Preheat MICROWAVE BROWNER at HIGH 2½ minutes.
2. Remove pizza from carton; remove plastic wrap.
3. Place frozen pizza on center of PREHEATED browner.
4. Cook at HIGH 5 to 6 minutes or until center cheese is melted, rotating BROWNER ½ turn after each 2 minutes of cooking.
*Microwave oven directions are based on test results using 600 to 700 watt counter top ovens.
TRY ALL EIGHT VARIETIES OF CELESTE PIZZA
• CHEESE • DELUXE
• PEPPERONI • SAUSAGE & MUSHROOM
• SAUSAGE • CANADIAN STYLE BACON
• SUPREMA • VEGETABLE
AVAILABLE IN TWO CONVENIENT SIZES!

Product: Celeste Pizza For One
Designers: Owen W. Coleman and Abe Segal
Design Firm: Coleman, LiPuma, Segal & Morrill Inc., New York, NY
Client: The Quaker Oats Company, Chicago, IL

The redesign of Celeste products for The Quaker Oats Company was initially developed utilizing computer graphics. A friendly, realistic painting of Mama Celeste within an oval plaque was introduced and resulted in an increase in sales.

Product: Cheese
Designer: Shining
Design Firm: Shining, France
Client: Gervais Danone, France

This plastic container houses whole milk cheese from Normandy.

Product: Vegetarian Lasagne
Designer: Mary Lewis
Design Firm: Lewis Moberly, London, England
Client: The Boots Company, Ltd., Nottingham
Award: 1986 Clio nomination, International Packaging

A new, contemporary image of the health-food community is brought about by a simple, appealing package.

Product:	Breakfast Biscuits
Designer:	Ennio Lucini
Design Firm:	Studio Elle, Milano, Italy
Client:	Galbusera for La Rinascente, Italy

This design for breakfast cookies, customarily eaten in Italy during the morning hours, was adopted to differentiate these Galbusera products from others in the line. The illustration refers to breakfast time.

Product: Colombo Yogurt
Designer: Group Four Design
Design Firm: Group Four Design, Avon, CT
Client: Colombo

Use of a licensed Smurf character brings great youth appeal to the packaging for childrens' yogurt.

Product: Pillsbury's Best Refrigerated Cookies
Designer: Kornick Lindsay
Design Firm: Kornick Lindsay, Chicago, IL
Client: Pillsbury

New graphics were developed to introduce a line of refrigerated cookies in a plastic film package. A single cookie was chosen to communicate each flavor, along with color-coordinated flavor names. Fine blue pin stripes on the opaque white background provide a touch of quality to the line.

Product: Dinner Classics
Designer: Wayne Krimston
Design Firm: Murrie White Drummond & Lienhart Associates, Chicago, IL
Client: Armour Food Company

This package helped establish a new price segment in the frozen dinner category. The package projects an elegant, quality image needed to support its premium price.

Product: Classic Lite
Designer: Thomas Q. White
Design Firm: Murrie White Drummond & Lienhart Associates, Chicago, IL
Client: Armour Food Company

Low calorie products traditionally have been packaged in white or pale colors. The rich color background of Classic Lite projects an image of high quality taste appeal.

Product: Welch's Products
Designer: Peterson Blyth Associates
Design Firm: Peterson Blyth Associates, New York, NY
Client: Welch Foods, Concord, MA

Welch's communicates its history of producing natural, family-oriented fruit products. Scrumptious fruit against a latticework background brings across Welch's tradition for high quality products.

Product: Classy Crisps
Designer: Amy Leppert
Design Firm: Murrie White Drummond & Lienhart Associates, Chicago, IL
Client: Beatrice

Classy Crisps packaging creates an upscale, contemporary brand identity with strong appetite appeal and high quality imagery.

Product: Rena Cakes
Designer: G.E.2A Bureau de Création Graphique
Design Firm: G.E.2A Bureau de Création Graphique, Lambersart, France
Client: Agence Fusion, Lille, France

A colorful array of packages represents the various types of fruit cakes made in France. The Scot cake is packaged with a plaid border, and the home cake package has a gold band with the Rena name imprinted in red.

Product: Milk Chocolate and Almond Bars
Designer: Thomas Q. White
Design Firm: Murrie White Drummond & Lienhart Associates, Chicago, IL
Client: Nestle Foods Corporation

Rich colors with silver accents and the bold Nestle logo, synonymous with high-quality chocolate, project an image of richness reflective of Nestle's European heritage.

Product: Andes Candies
Designer: Dickens Design Group
Design Firm: Dickens Design Group, Chicago, IL
Client: Andes Candies, Delavan, WI

This new package idea shows fresh individual wafer bars and creates an exciting small gift package for a growing variety of chocolate candy wafers.

Product: Frango Chocolates
Designer: Wayne Krimston
Design Firm: Murrie White Drummond & Lienhart Associates, Chicago, IL
Client: Marshall Field's

An elegant logotype and rich, flavor-coded background colors were used for these Frango products.

Product: Mott's Brik Pak
Designers: Ronald Peterson and Barbara Wentz
Design Firm: Peterson & Blyth Associates, Inc., New York, NY
Client: Mott's

Peterson & Blyth Associates was one of the first design firms to put aseptic brik paks in a dramatic outer sleeve for stronger shelf impact. The packaging's playful graphics appeal to the children and teenagers that this new product is positioned for and emphasize the product's wholesome nature as well as the Mott's brand name.

Product: Breakfast Yogurt
Designers: Sen Gupta and Nicole Cuomo
Design Firm: Shining, France
Client: Gervais Danone, France

This new yogurt mixed with dried fruits and cereals was to be positioned as a breakfast item. The containers were designed so that the product could be stored in the refrigerator to stay fresh.

Product: Land O'Lakes Spoonery Cheese
Designer: Peterson & Blyth Associates, Inc.
Design Firm: Peterson & Blyth Associates, New York, NY
Client: Land O'Lakes

Land O'Lakes cut its packaging costs considerably by switching from a glass crock to this plastic tub container. The form-fill-seal tubs are more economical to produce and to ship.

Product: International Dessert Cheesecakes
Designer: Thomas Q. White
Design Firm: Murrie White Drummond & Lienhart Associates, Chicago, IL
Client: Kitchens of Sara Lee Inc.

Dramatic photographs against a white background generate an image of elegance and taste appeal for Sara Lee.

Product: Marshall Field's Gourmet
Designer: Wayne Krimston
Design Firm: Murrie White Drummond & Lienhart Associates, Chicago, IL
Client: Marshall Field's

Field's specialty foods incorporated a new "Gourmet" name and "Bountiful Basket" symbol to achieve an impression of quality.

Product: Chinese Sauces
Designer: Mary Lewis
Design Firm: Lewis Moberly, London, England
Client: Spillers Foods, Ltd., Surrey, England

A black photographic background creates a striking and atmospheric pack with bright vignettes.

Product: Ready-to-Eat Canned Beef
Designer: Gió Rossi
Design Firm: Image Plan International, Milano, Italy
Client: Trinity, Milano, Italy

The new, upgraded package for Manzotin includes a peel-off lid, a simple label, and a clean illustration.

Product: Dundee Marmalade
Designers: Owen W. Coleman and John Rutig
Design Firm: Coleman, LiPuma, Segal & Morrill Inc., New York, NY
Client: Nestle Foods Corporation, White Plains, NY

As a substitute for producing original ceramic containers, Nestle settled on this upgraded design image for Dundee products. The paper label reflects the longevity of the Dundee product line.

Product: Stokely Canned Foods
Designer: Larry Riddell
Design Firm: Gerstman & Meyers Inc., New York, NY
Client: Stokely USA Inc., Indianapolis, IN

This simple design has eliminated the previous clutter associated with the brand, and has pulled Stokely away from the "sea of look-alikes."

Product: Vlasic Pickles
Designer: Michael Kelly
Design Firm: Murrie White Drummond & Lienhart Associates, Chicago, IL
Client: Vlasic Foods Inc.

Color-coding, new flavor descriptions and the introduction of an innovative "seasoning scale" help consumers identify preferred tastes among the many varieties in the Vlasic Foods line.

KOSHER
Crunchy Dills
Old Fashioned, Classic Dill Taste
LIGHTLY SEASONED 1 2 3 4 HIGHLY SEASONED
TM

Chapter 2

Beverages

If there can be news in beverage packaging, it is that manufacturers continue to produce beverages in single-serving sizes. This trend, which is most prevalent among juices, wines, and wine coolers, has led to the packing of single-serving bottles and cans in three-packs, four-packs and six-packs.

Because of the limited space on single-serving containers, label designs must be neater and more organized, especially if a large amount of product information is to be presented. This sort of design problem lends itself to color coding and other devices that distinguish or identify the product in some way without using up valuable label space.

One other trend is also apparent in current beverage packaging: a move toward more refined designs, particularly in European packaging. Of particular note are the sophisticated designs of Régilait powdered milk (p. 101), Misura low-calorie soft drinks (p. 102), and Percol Night and Day coffee (p. 106).

The current practice of producing single-serving containers and packing them in three-packs, four-packs, and six-packs originated in the beer and soft-drink industries with the traditional six-pack. The limited label space on single-serving cans and bottles requires neat, well-organized graphics such as those on these bottles of The Original Irish Cork Stout. Created by Mary Lewis of Lewis Moberly, the design carries an air of stateliness and nostalgia.

The graphic design for Misura low-calorie soft drinks is one of the most sophisticated designs in the current marketplace. Created by Design Group Italia, this design uses colored bands of varying widths to identify the various flavors and create a bright display on the store shelf. Both the logo type and the typography are quite progressive and project a contemporary image.

Product: Albastella Table Wine
Designer: Gió Rossi
Design Firm: Image Plan International, Milano, Italy
Client: Noé/Cavit Cellars, Italy

A beautiful illustration of wildlife contributes to a feeling of festivity and freshness.

Product: Blanton Kentucky Straight Bourbon Whiskey
Designers: Bill Reidell and Adam D'Addario
Design Firm: The Creative Source Inc., New York, NY
Client: Blanton Distilling Company
Awards: 1985 Clio Award finalist; 1985 PDC Gold Award finalist

A corked decanter with a parchment label represents this premium-priced bourbon. Each label carries hand-written notes that give information such as the date and warehouse destination of the individual bourbons.

Product:	Champagne
Designer:	Mary Lewis
Design Firm:	Lewis Moberly, London, England
Client:	Asda, Leeds, England

The classic, traditional style of this packaging enables it to reflect quality in a modern market.

Product: Inglenook Carafes
Designers: Dante Calise, Karen Vaccaro, and Diamond-Bathurst
Design Firm: The Creative Source Inc., New York, NY
Client: Heublein Inc.

A new glass carafe design inspired the cathedral-shaped labels for Vin Rosé, Chablis, and Burgundy wines. This represents Inglenook's first entry into carafes, designed to graciously serve wines at the table.

Product: Sunflo Juices
Designer: Selame Design
Design Firm: Selame Design, Newton Lower Falls, MA
Client: Sunflo

Drawings of stylized graphic fruits are used to set Sunflo juices apart from others on the shelf.

Product: James Paine Ales
Designer: Peter Windett & Associates
Design Firm: Peter Windett & Associates, London, England
Client: James Paine Brewery, Ltd.

All the various James Paine Ales, like the Brown, Pale and Dark Ales, are bottled with a general design pattern that is in keeping with each different ale.

Product: Normandy Cider
Designer: Mary Lewis
Design Firm: Lewis Moberly, London, England
Client: Tesco, Herts, England

This represents a unique packaging idea for a highly traditional, all-natural cider. The four woodcuts depict traditional cider-making.

Product: Athletic Performance Drink
Designer: Kathleen Campbell
Design Firm: Tom Campbell & Associates, Inc., Los Angeles, CA
Client: Vitex Foods, Los Angeles, CA
Awards: 1985 New York Art Directors' Club award; 1985 *Design Review* Designer's Choice Award; 1985 Los Angeles Art Directors' Club award

Hi-tech graphics and a bright recognizable logotype create credibility to appeal to serious athletes.

Product: Peaches 'N Cream
Designers: Thomas D'Addario and Linda Chow
Design Firm: The Creative Source Inc., New York, NY
Client: Heublein Inc.

Heublein Peaches 'N Cream Liqueur is packaged in an appealing gift box with a four-panel, soft illustration of peaches and cream.

Product: Light Ale
Designer: Mary Lewis
Design Firm: Lewis Moberly, London, England
Client: Tesco, Herts, England

This simple, yet bold design for Light Ale creates a strong shelf display. Price alone is the key positioning factor.

Product: Lemonade and Ice Tea
Designer: Dixon & Parcels Associates, Inc.
Design Firm: Dixon & Parcels Associates, Inc., New York, NY
Client: 4C Foods Corporation, Brooklyn, NY

Brand identification and product type are clearly visible in shelf displays, and a thirst-quenching image is projected to consumers.

Product: Caffe Sport Liqueur
Designer: Carpano in-house designers
Design Firm: Carpano
Client: Carpano, Torino, Italy

An eye-catching, bright label creates high visibility for this liqueur.

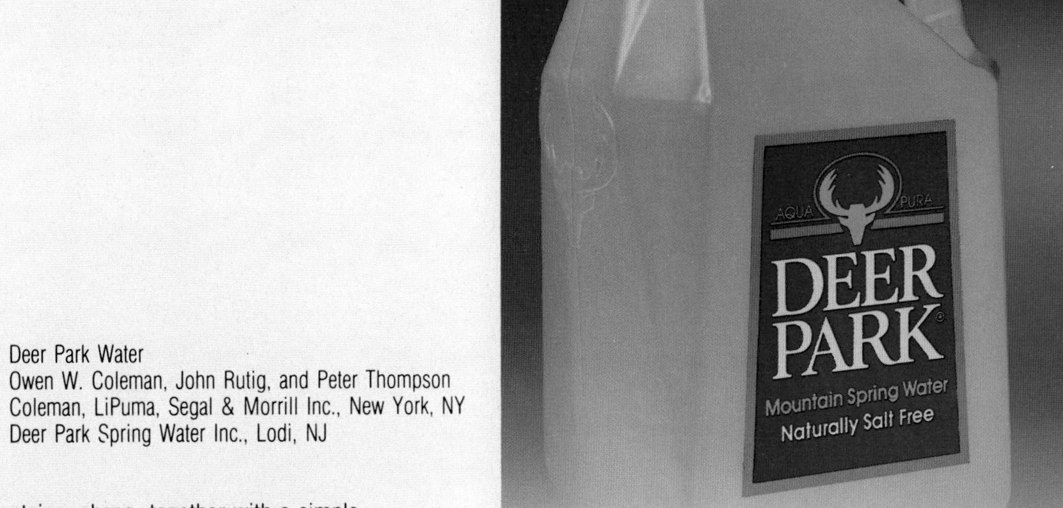

Product: Deer Park Water
Designers: Owen W. Coleman, John Rutig, and Peter Thompson
Design Firm: Coleman, LiPuma, Segal & Morrill Inc., New York, NY
Client: Deer Park Spring Water Inc., Lodi, NJ

This new container shape, together with a simple bold label and emblem for Deer Park Spring Water, was carried across all Deer Park signage and corporate identification.

Product: Sunkist Natural
Designer: Jacquie Macconnell Fauter
Design Firm: Peterson Blyth Associates, New York, NY
Client: Delmonte Franchise, Atlanta, GA

This new product features the same logo that appears on Sunkist soda, but presents it in orange here. The vibrant green background color was selected to communicate the product's natural ingredients.

Product: Dalesman Best Bitter
Designer: Mary Lewis
Design Firm: Lewis Moberly, London, England
Client: Tesco, Herts, England
Award: Clio Gold International Packaging, 1986

The packaging for this product—which comes from a small, traditional brewer in the Yorkshire dales—is designed to emphasize quality, heritage, and authenticity.

Product:	Original Irish Cork Stout
Designer:	Mary Lewis
Design Firm:	Lewis Moberly, London, England
Client:	Tesco, Herts, England
Award:	Clio Gold International Packaging, 1985.

An old-time label design, a nostalgic black-and-white illustration, and an old-fashioned bottle shape help communicate that this package contains "the original cork stout."

Product:	Rocay Amat Champagne
Designer:	Gió Rossi
Design Firm:	Image Plan International, S.p.A., Milan, Italy
Client:	San Sadurnide Noya, Spain

A foil-neck champagne bottle with a vertical ribbon of gold and maroon running through the traditional gold-laced labels communicates a sense of quality and long-time establishment.

Product:	Proprietors Selection Champagne
Designer:	Coming Attractions Communications Services
Design Firm:	Coming Attractions Communications Services, San Francisco, CA
Client:	Spectrum Foods, San Francisco. CA

To complement the atmosphere of their fine restaurants, the owners of Spectrum Foods requested a proprietary champagne package. The key to this design is the use of fine art as the primary graphic element.

Product: San Pellegrino Soft Drinks
Designer: Gió Rossi
Design Firm: Image Plan International, Milano, Italy
Client: San Pellegrino, Milano, Italy

A bright star represents a wide line of San Pellegrino soft drinks. Colors on bottles and cans signify flavors.

Product: Davide Campari
Designer: Davide Bolzonella
Design Firm: G & R Associati, Milan, Italy
Client: Davide Campari, Milan, Italy

A brilliant gift box houses the best of Davide Campari spirits.

Product:	Port Wine
Designer:	Lewis Moberly
Design Firm:	Lewis Moberly, London, England
Client:	Asda, Leeds, England

A sophisticated package helps position this brand of port in a modern market and, at the same time, retains the traditional image of the product.

Product: Aberlour Scotch Whiskey
Designer: Sopha
Design Firm: Sopha, Paris, France
Client: Pernod SA, Paris, France

This is a masculine, chic package for the Pernod Scotch bottle.

Product: Eloise Villacosta White Wine
Designer: Gió Rossi
Design Firm: Image Plan International, S.p.A., Milan, Italy
Client: Villacosta, S.p.A., Italy

Available in this rich green and gold package, this Villacosta white wine product is bottled in Italy.

Product: Pina Colada
Designers: Dante Calise, Karen Vaccaro, and Thomas D'Addario
Design Firm: The Creative Source Inc., New York, NY
Client: Heublein Inc.

When Heublein reformulated its Pina Colada Mixed Drink, the company decided to redesign the product label to convey the theme of "all natural flavorings." The photography on the label accomplishes this by presenting appetizing fruits in a tropical setting.

Product: Davide Campari
Designer: Davide Bolzonella
Design Firm: G & R Associati
Client: Davide Campari, Milano, Italy

These Campari liquors and spirits are packaged in a box that creates a unified graphic image when closed.

Product: Decaffeinated Coffee
Designer: Groupe Design MBD
Design Firm: Groupe Design MBD, Paris, France
Client: Jaques Vabre, France

The illustration matches the product name, "Night and Day," and presents an image associated with decaffeinated coffee.

Product: Powdered Milk
Designer: Hot Shop
Design Firm: Hot Shop, Paris, France
Client: France Lait, France

A colorful sketch links the product to the powdered milk category.

Product: Printanet Wine
Designer: Hotshop
Design Firm: Hotshop, Paris, France
Client: S.B.V. Chantovent, France

Here, a new approach to wine packaging makes it easier for the manufacturer to ship the product and for the consumer to store it. This light red wine has caught up with the pace of contemporary life by catering to "people on the go."

Product: Inglenook 1.5 Liter with On-Pack
Designers: Thomas D'Addario, Linda Chow, and Karen Vaccaro
Design Firm: The Creative Source Inc., New York, NY
Client: Heublein Inc.

Inglenook Vineyards offered 187 ml trial-size bottles as on-packs to the regular 1.5 liter bottles, thus creating an industry first.

Product: Low Calorie Soft Drinks
Designer: Design Group Italia
Design Firm: Design Group Italia, Milano, Italy
Client: Plada, Italy

These "no-sugar-added" drinks were the first to appear on the Italian market. Color bands differentiate the various flavors and create a bright display.

Product: Baybry's Champagne Cooler
Designer: Obata Design
Design Firm: Obata Design, St. Louis, MI
Client: Anheuser-Busch Companies, St. Louis, MI

A simulated champagne stopper, printed foil neck wrap, and rich color are employed to create an impression of quality.

Product: Lait Gervais
Designers: Sen Gupta and Evelyn Delassus
Design Firm: Shining, Paris, France
Client: Laiterie de Villecomtal, Aveyron, France

Graphics convey an authentic country look and a wholesome quality for Gervais sterilized milk.

Product: Arrow Schnapps and Liqueur
Designers: Thomas D'Addario, Dante Calise, and Karen Vaccaro
Design Firm: The Creative Source Inc., New York, NY
Client: Heublein Inc.

The expanding Arrow cordial line required segmentation. Therefore, the fruit-flavored Liqueur and Schnapps labels were designed with special borders to differentiate them.

Product: Squirt Plus
Designer: Thomas Q. White
Design Firm: Murrie White Drummond & Lienhart Associates, Chicago, IL
Client: Squirtco

The package graphics for this first "multi-vitamin enriched" soft drink create an all-natural image.

Product: Merchant Vintners Wine
Designer: David Hillman
Design Firm: Pentagram Design Ltd.
Client: Merchant Vintners

The label design for this Italian wine is based on the red, white, and green of Italy's national flag.

Product: Night and Day Coffee
Designer: Mary Lewis
Design Firm: Lewis Moberly, London, England
Client: Food Brands Group Ltd., England

This first low caffeine, rather than *no* caffeine, product on the market was designed to make a new statement in the traditional coffee market. The packaging is aimed at the health-conscious individual who still wants the taste and quality of real coffee, but with less caffeine.

Product: Peach Schnapps
Designer: Frederick Mittleman Designs
Design Firm: Frederick Mittleman Designs, New York, NY
Client: National Distillers Products Company, New York, NY

This illustrative design conveys a strong taste appeal for DeKuyper Peach Schnapps. It also creates a strong shelf presence.

Product: Nano Sparkling White Wine
Designer: Gio´Rossi
Design Firm: Image Plan International, Milano, Italy
Client: San Pellegrino, Milano, Italy

Nano, meaning dwarf, is the name of this sparkling white wine, which is bottled with metal caps instead of corks.

Product: Arrow Schnapps
Designers: Thomas D'Addario and Dante Calise
Design Firm: The Creative Source Inc., New York, NY
Client: Heublein Inc.

The distinctive type styles, hand lettering, and subject-related background colors on these labels represent a new marketing strategy for Arrow Schnapps.

Chapter 3

HEALTH AND BEAUTY AIDS

A divergence appears to be occurring in the packaging design for health and beauty aids, the result of three very different marketing approaches.

One approach, which seeks to project a medical image, has brought about an extremely straightforward and sometimes austere type of design. American packaging of this type tends to be particulary severe. The package for the AVI/3M IV infusion device (p. 123), for instance, positions unobtrusive type, which serves merely as a labeling device, against a medicinal white background. Medically oriented design from Europe is also quite straightforward but tends to use color and simple but sophisticated graphics.

Another marketing approach seeks to promote the natural aspects of a product and results in package designs that convey the goodness of nature. Consider the subtly-colored illustrations of flowers on the labels of Yves Rocher natural shampoos (p. 116) or the repeated illustration of a bee positioned against a soft yellow background on the label of Perlier skin-care cream.

The third marketing approach apparent in the current marketplace seeks to create a rich, sophisticated image and is particularly prevalent among men's products. The packaging for Kelemata Cologne (p. 114) is a powerful and elegant example of this—it features a stong black bottle, a deep blue label with sophisticated type treatment, and a deep blue and black outer box with type that corresponds to that on the label.

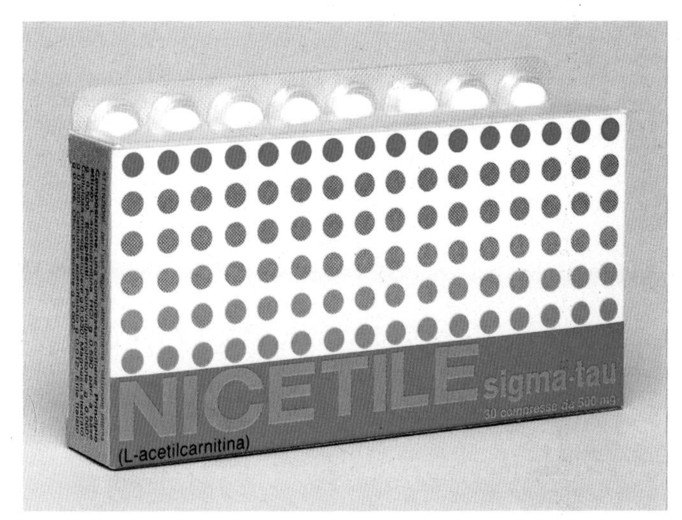

This simple, but sophisticated box, which contains a Sigma-Tau prescription drug, is a well-designed example of medically oriented packaging in Europe. In the U.S., the medical approach to marketing health and beauty aids has led to a more severe, often monochromatic type of design.

A delicate illustration of soft pink flowers and a gentle blue bird conveys all the goodness of nature. In choosing a natural approach for the package design of its Magnolia Soap, Scarborough and Company has associated its product with freshness, purity, and fragrance.

The designers of the L'Envie Parfume Shampoo bottle have created a rich, elegant image for this shampoo. The understated background color, the sophisticated type treatment, and the stylized symbols for fragrance identification all contribute to the effect.

Product: Roberts Talcum Powder, Baby Care Products
Designer: Gio´ Rossi
Design Firm: Image Plan International, Milano, Italy
Client: Manetti L. Roberts H. & C. S.p.A., Firenze, Italy

This talcum powder container has been updated, but the original design has basically remained the same. The entire line of baby care products is designed for safety around young children, as demonstrated by the rounded edges.

Product: Pantene for Men
Designer: Heide Mohrmann Tidwell
Design Firm: Peterson Blyth Associates, New York, NY
Client: Richardson-Vicks, Wilton, CT

A foil stamped "M" rendered in bold silver brushstroke visually communicates that the products are for men. The masculine image is further reinforced by use of a deep shade of gray.

Product: Terme di Saturnia Cosmetics
Designer: Nadia Poggi
Design Firm: Essevi Sales Promotion, Milano, Italy
Client: Elebel S.p.A., Milano, Italy

Saturnia, in southern Italy, is known for its thermal baths. Terme di Saturnia is a line of high-quality cosmetics, packaged in practical, contemporary containers.

Product: A.C.T.
Designer: Wayne Krimston
Design Firm: Murrie White Drummond & Lienhart Associates, Chicago, IL
Client: Dena Vic-Meta Henna International

Bold primary colors create a dynamic international look and position the line for use by swimmers.

Product: Beeswax Cream
Designer: Staff, Kelemata
Client: Kelemata, Torino, Italy

This cream consists of all natural ingredients, as do the other products in this skin-care line. The jars inform the consumer of this information, an important aspect of Kelemata's marketing strategy.

Product: Phytosolba Hair Products
Designer: Staff designers, Phytosolba
Client: Phytosolba, Paris, France

The packaging design for these hair products attractively indicates the primary ingredients, plant extracts.

Product: Kelemata Cologne
Designer: Staff, Kelemata
Client: Kelemata, Torino, Italy

The black bottle and the larger type on the box distinguish this men's product from Kelemata's products for women.

Product: Kelemata Eye Contours
Designer: Staff, Kelamata
Client: Kelemata, Torino, Italy

A clean, uniform design was chosen for these eye-contour products manufactured by Kelemata.

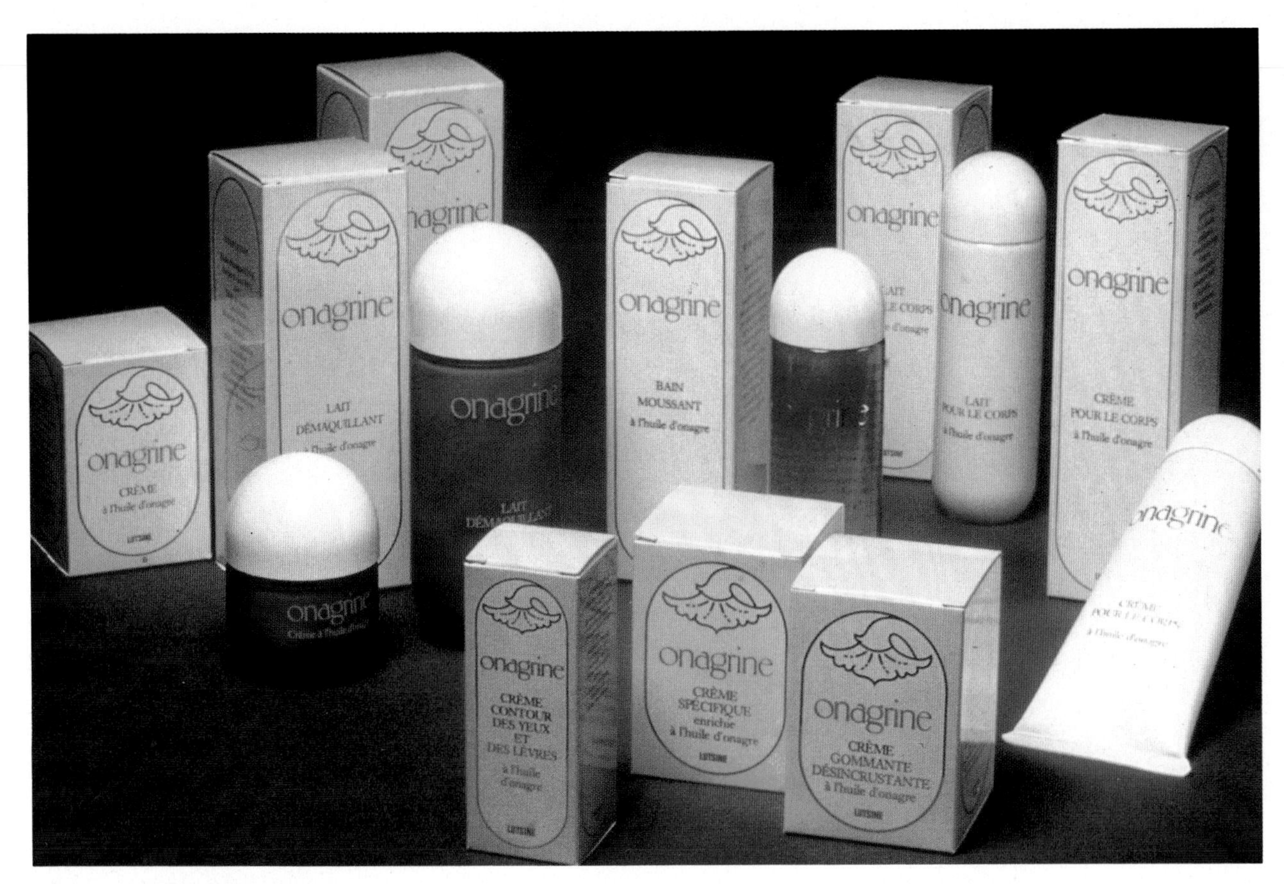

Product: Onagrine Skin Products
Designer: Sylvie Lecot
Design Firm: Lonsdale Design & Communication, Paris, France
Client: Laboratories Lutsia, Paris, France
Awards: Annik de Cholet—Antoni

This delicate design was created for a line of high-quality skin care products.

Product:	Natural Shampoos
Designer:	Hotshop
Design Firm:	Hotshop, Paris, France
Client:	Yves Rocher, Paris, France

Convenient plastic bottles, safe for use in the bathroom, represent a simple design idea.

Product: Lady Stetson Perfume
Designer: Group Four Design
Design Firm: Group Four Design, Avon, CT
Client: Pfizer/Coty

Gold typography complements the perfume color in this new bottle design for Lady Stetson.

Product: L'Envie Parfum
Designers: Heide Mohrmann Tidwell, and Marianne Walther
Design Firm: Peterson Blyth Associates, New York, NY
Client: S.C. Johnson & Son, Racine, WI

The classically-shaped bottle's flat front surface gives this new product greater impact on the supermarket or drugstore shelf. The 12-oz. bottle has a twist-off, snap-on cap, which reinforces the brand's unique qualities.

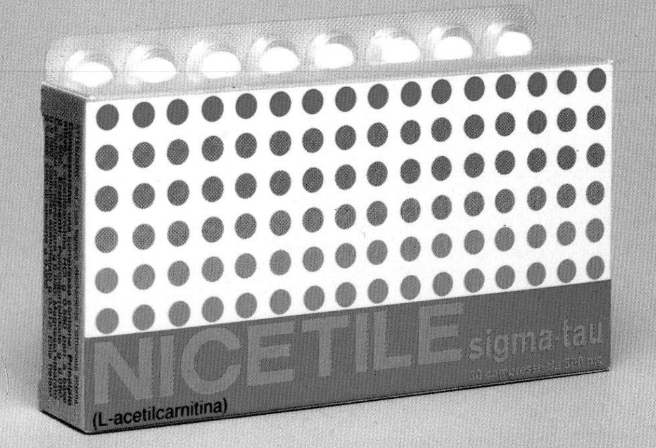
NICETILE
sigma-tau
(L-acetilcarnitina)

sigma-tau
DEB
20 capsule
DEBRIDAT®
Flacone da
152,5 g di granulare
DEBRIDAT®

Product: Sigma Tau Prescription Drugs
Designer: Gabriele Stocchi
Design Firm: Staff, Sigma Tau
Client: Sigma Tau industrie farmaceutiche riunite S.p.A., Rome, Italy

This line of various pharmaceutical products is graphically unified in terms of packaging. Soft coloring is relaxing to the eye and suggests relief.

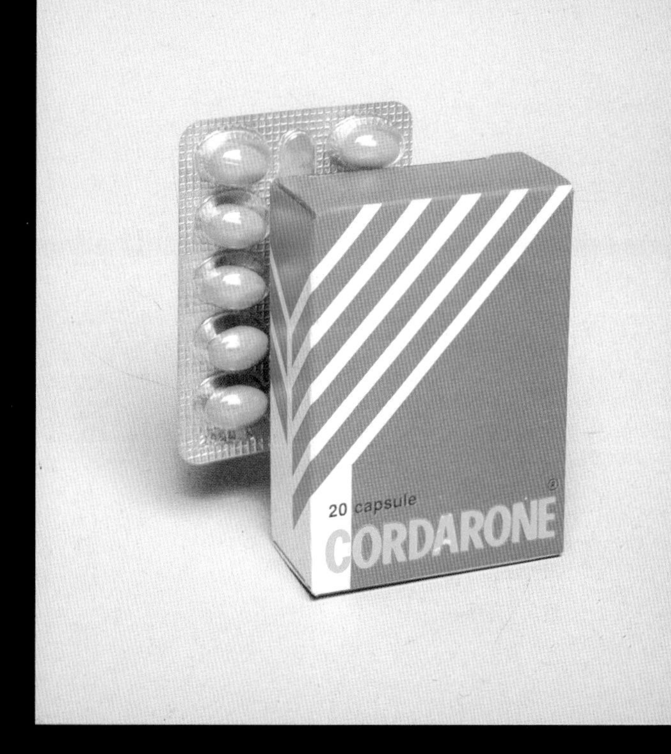

Product: Aquafresh Toothpaste
Designer: Gió Rossi
Design Firm: Image Plan International, Milano, Italy
Client: Beecham , Milan, Italy

Through the years, Aquafresh has found success with its mainly green package and popular illustration of a swirl of toothpaste on a toothbrush. The water in the background suggests cool, minty flavor, as does the green on the package.

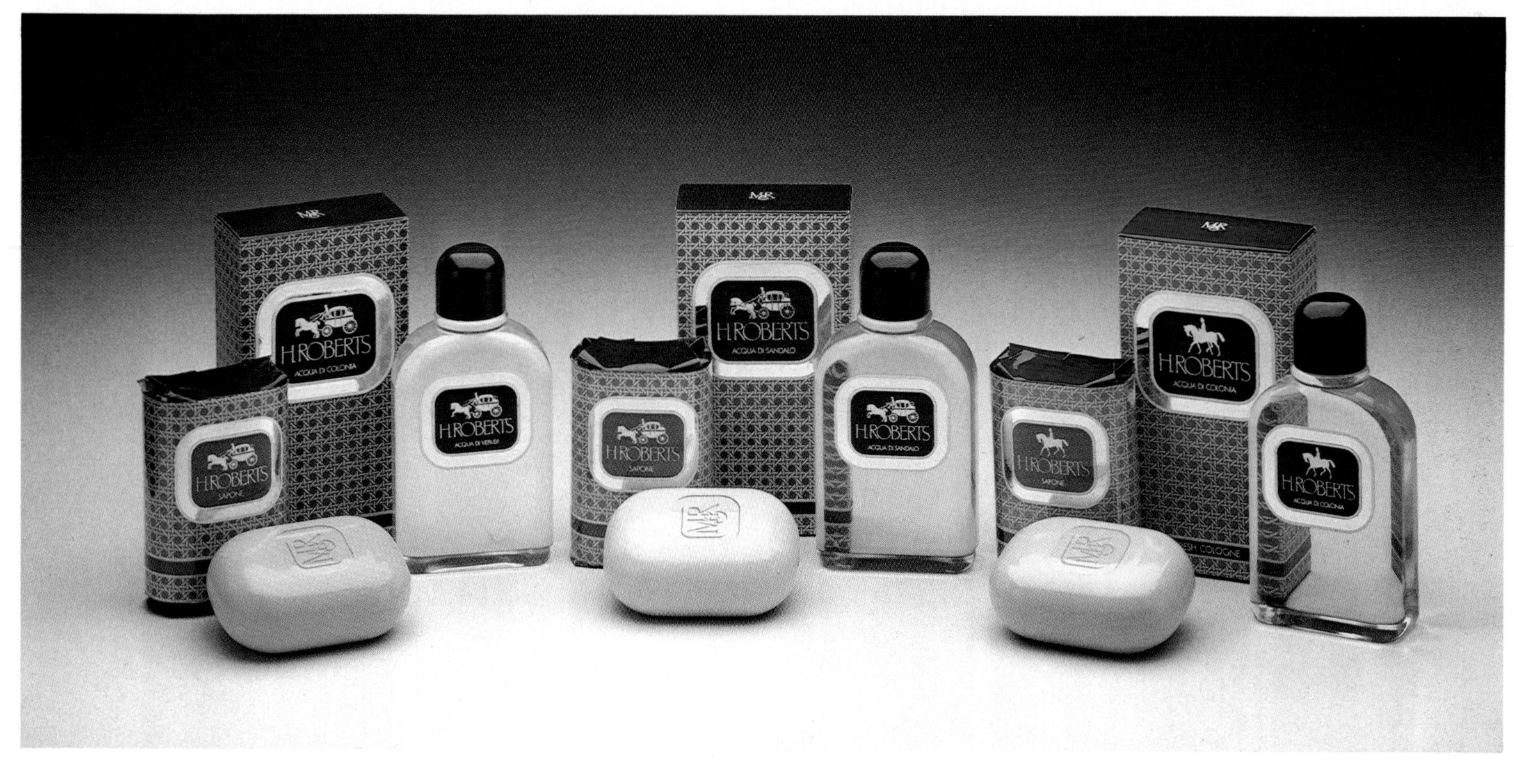

Product: Colognes and Soaps
Designer: Gió Rossi
Design Firm: Image Plan International, S.p.A., Milan, Italy
Client: Manetti L. Roberts, Firenze, Italy

Roberts appeals to the more traditional sense in their male target audience, as seen by the packaging designs for their colognes and soaps.

Product: Badedas Bath Products
Designer: Gió Rossi
Design Firm: Image Plan International, S.p.A., Milan, Italy
Client: Badedas, Italy

Various Badedas bath products are packaged similarly in green and yellow durable plastic containers, safe for use in the bathroom.

Product: Men's Toiletries
Designer: Gió Rossi
Design Firm: Image Plan International, S.p.A., Milan, Italy
Client: Atkinsons-Unilever Group, Milan, Italy

A masculine image, created by a gold-on-black label, the "Executive" product name, and the product's logo, sets these various men's products apart from feminine toiletries.

Product: English Lavender and Gold Medal Bath and Shower Products
Designer: Gió Rossi
Design Firm: Image Plan International, S.p.A., Milan, Italy
Client: Atkinsons-Unilever Group, Milan, Italy

A simple, handy container for these English Lavender and Gold Medal colognes and soaps is easy to hang in a bath or a shower. The boxes are decorative enough to give as gifts.

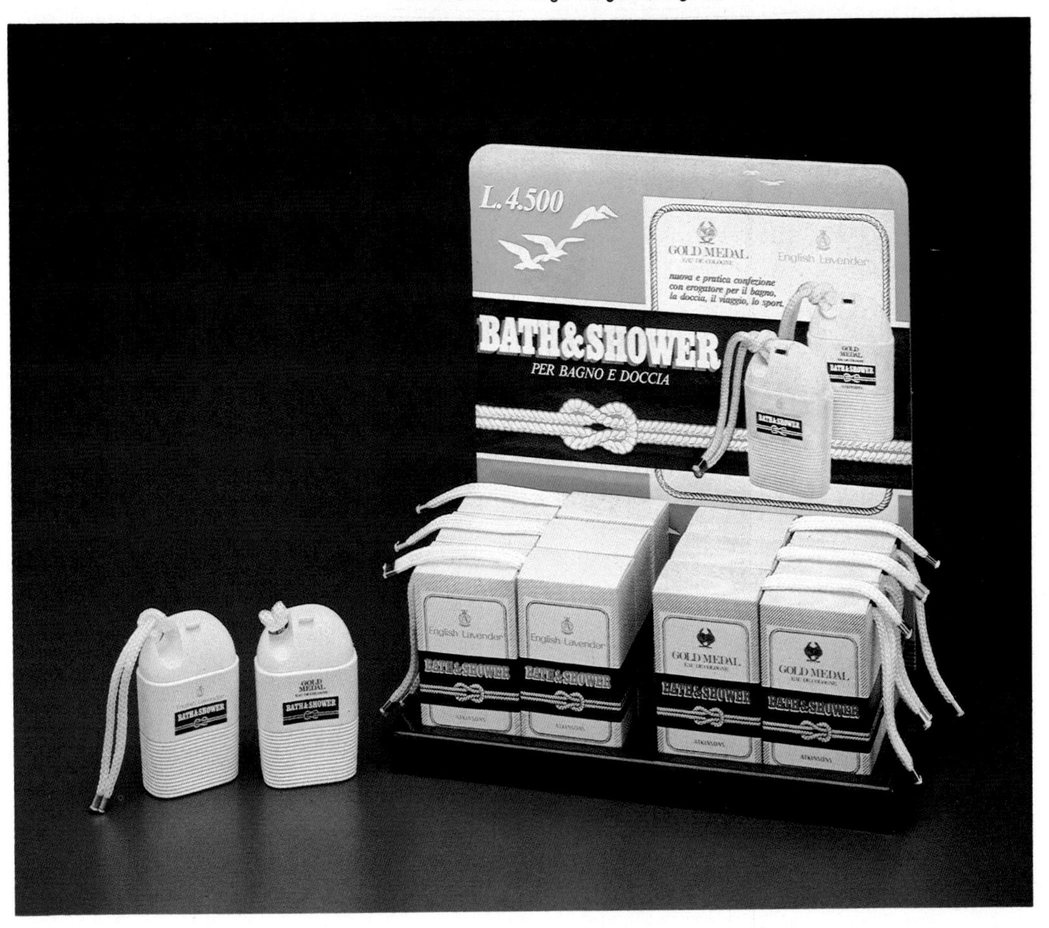

Product: The Dry Look
Designers: Edward Morrill and Simon Wong
Design Firm: Coleman, LiPuma, Segal & Morrill Inc., New York, NY
Client: The Gillette Company, Boston, MA

A unified, strongly masculine image was created for a line of hair care products for men. New hair care product forms such as the Styling Mousse were incorporated into the line.

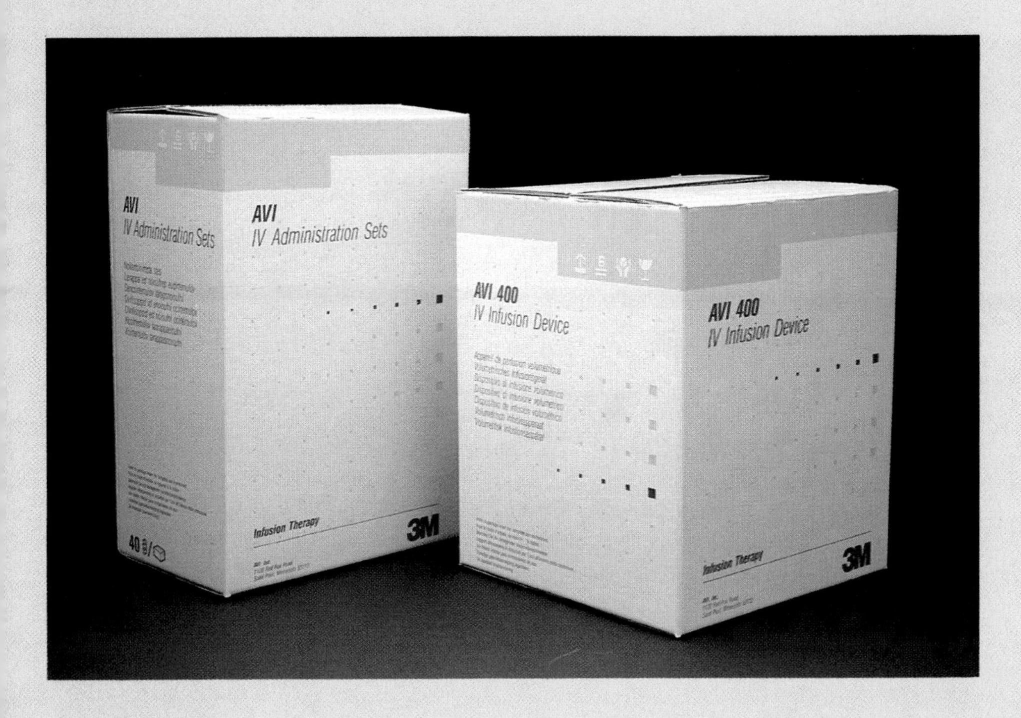

Product: AVI/3M
Designer: Aimee Hucek
Design Firm: Seitz Yamamoto Moss Inc., Minneapolis, MN
Client: AVI Inc./3M, St. Paul, MN

This package design was created to position AVI/3M Infusion Therapy in the international market as well as the domestic marketplace.

Product: French Line
Designer: Jöel Desgrippes
Design Firm: Desgrippes-Beauchant-Gobé-Cato, Paris, France
Client: Revillon, France

Men's products are packaged in masculine, dark containers with bold red and silver markings.

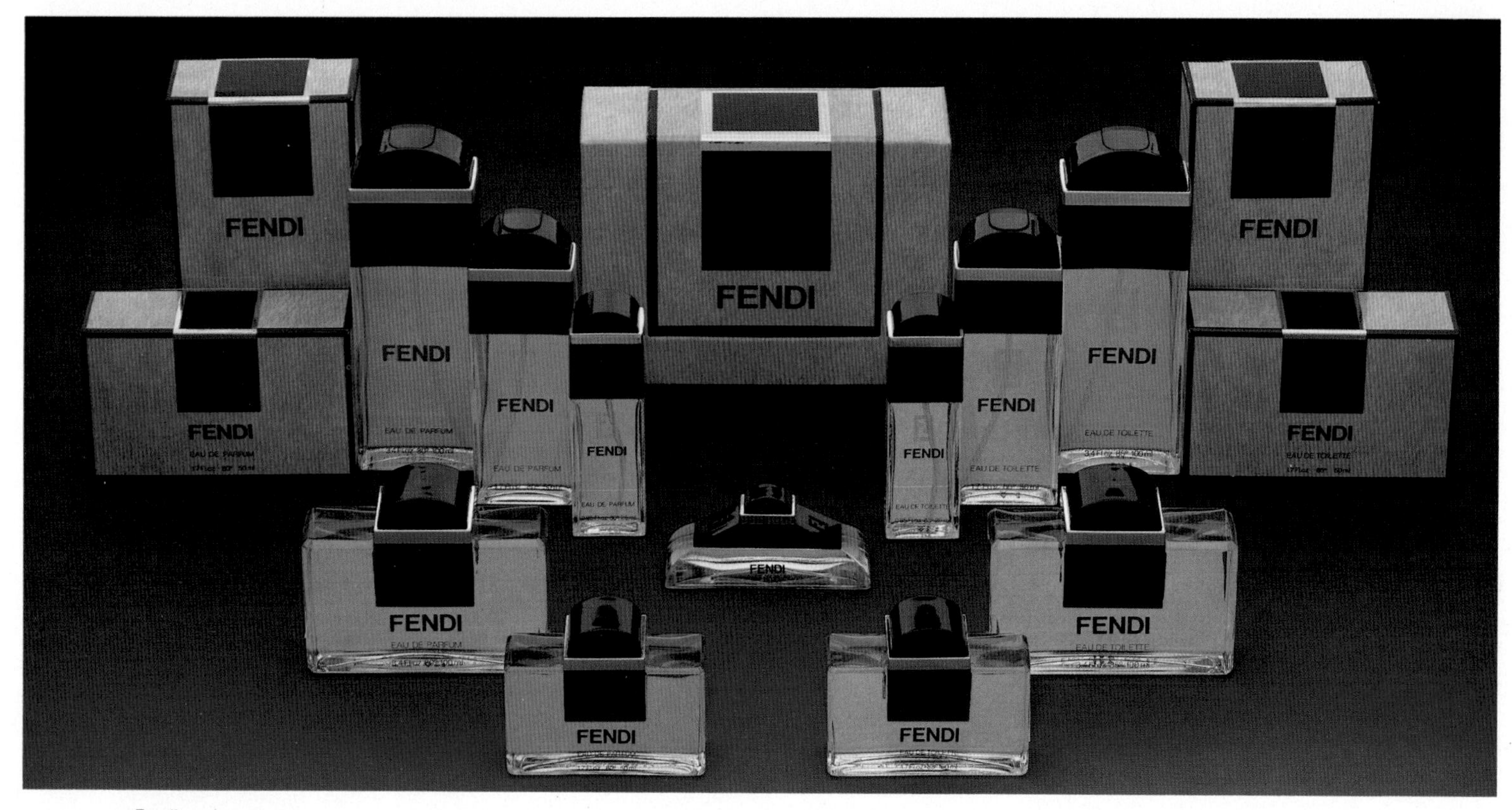

Product: Fendi
Designer: Karl Lagerfeld, for Fendi
Design Firm: Fendi Profumi, Parma, Italy
Client: Fendi Profumi, Parma, Italy

This elegant design for a bottle as well as a box includes the Fendi trademark embossed on each side of the cap.

Product: Kelemata Bath Products
Designer: Staff, Kelemata
Client: Kelemata, Torino, Italy

Here, bath foams are packaged in plastic containers and labelled with different colors to indicate foam type. Mint is coded green, rose is pink, and so on.

Product: Magnolia Soap
Designer: Peter Windett & Associates
Design Firm: Peter Windett & Associates, London, England
Client: Scarborough & Co.

A delicate pattern using soft pastels creates an aromatic feeling for these soaps. The soft illustration symbolizes freshness, purity, and fragrance.

Product: Rujel Cosmetics
Designer: Design Group Italia
Design Firm: Design Group Italia, Milano, Italy
Client: Gillette Italy

This sleek new design stands out on the shelf and makes it easy for both the consumer and the manufacturer to pack the product.

Product: Kleenex
Designers: Peter Petronio and Xavier de Bascher
Design Firm: Concept Groupe, Paris, France
Client: Sopalin, France

The idea behind this redesigning of Kleenex packaging was to cater more to both men and women, instead of only women.

Product: Tactics Men's Cosmetics
Designer: Mervyn Kurlansky
Design Firm: Pentagram Design Ltd.
Client: Shiseido Ltd., Japan

Tactics, a line of men's cosmetics, is manufactured by Shiseido, a very large Japanese company. The design concepts were aimed at an international market, although the product sells mainly in Japan. Boxes with the symbol printed in gold were distinguished by different-colored interiors that contrast with the simple white containers.

Product:	Bizarre Make-Up and Eau de Toilette
Designer:	Gió Rossi
Design Firm:	Image Plan International, Milano, Italy
Client:	Atkinsons — Unilever Group, Milano, Italy

Packaging for this extensive line of Bizarre make-ups is uniform and is clearly aimed for a young target audience. The flower on each package represents a product for young girls.

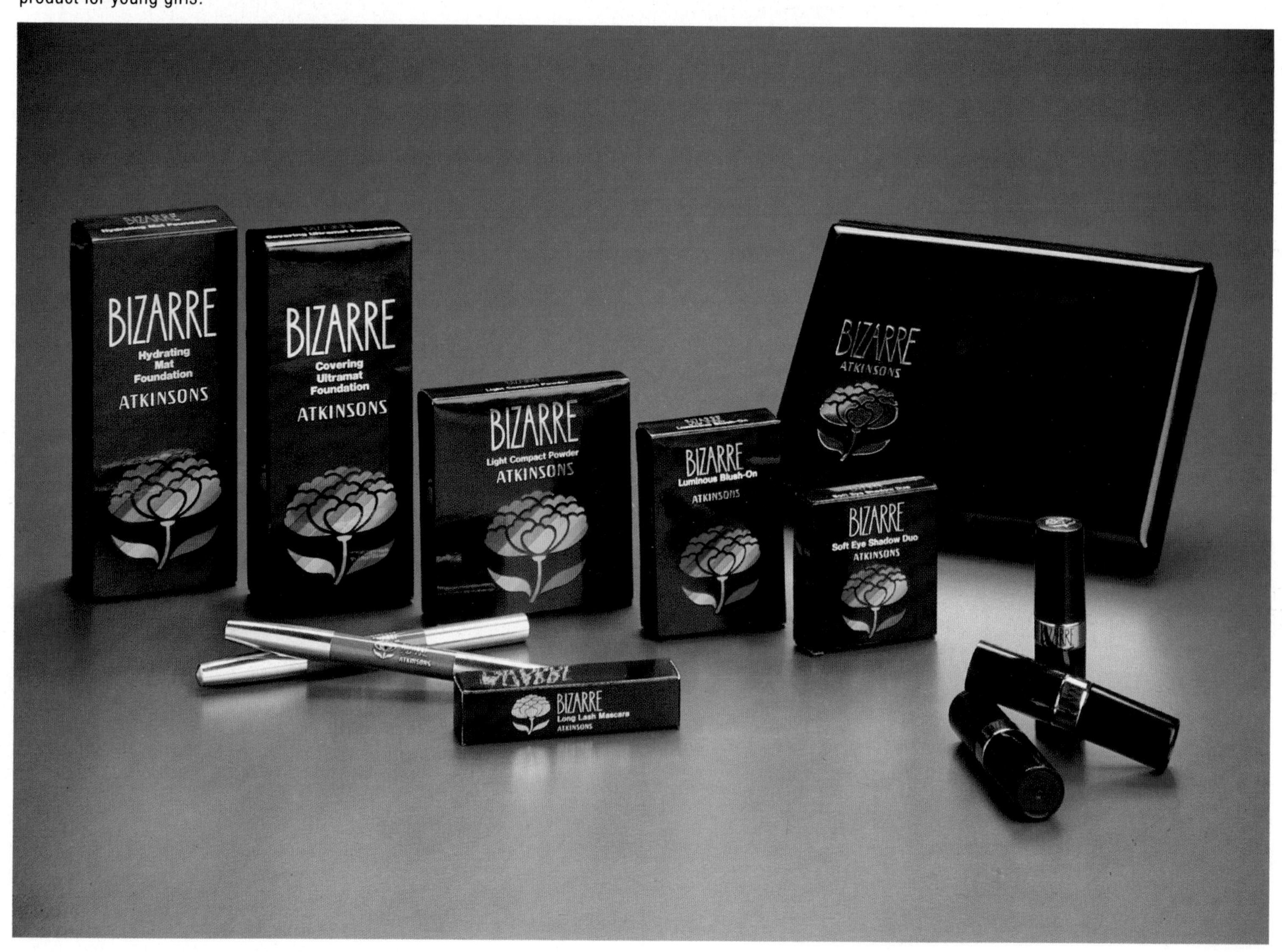

Product: Wella Shampoo
Designer: Design Group Italia
Design Firm: Design Group Italia, Milano, Italy
Client: Wella

A different bottle shape sets Wella Shampoo apart from competition. The waves on the cap repeat the Wella logo.

Product: Barbasol Glide Stick
Designer: Edward Morrill
Design Firm: Coleman, LiPuma, Segal & Morrill Inc., New York, NY
Client: Pfizer Inc., New York, NY

A new image for a line of Barbasol Glide Stick Deodorant products reflects a strong family tie-in with other Barbasol products.

Product: Sagamore
Designer: Löel Desgrippes
Design Firm: Desgrippes-Beauchant-Gobé-Cato, Paris, France
Client: Lancóme, France

Sleek containers, either black with silver or black with gold, enhance this men's cologne.

Product: Dimetapp
Designers: Owen W. Coleman and Abe Segal
Design Firm: Coleman, LiPuma, Segal & Morrill Inc., New York, NY
Client: A.H. Robins Company, Richmond, VA

A uniform design was created for a line of drug products available in both pill and liquid form. Bright coloring, which makes the package appealing, is effective for shelf display.

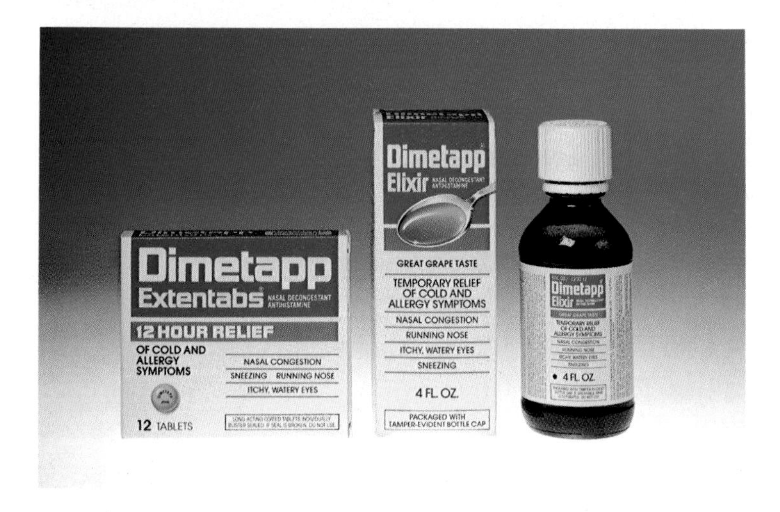

Product: Desitin Warm Relief
Designers: Edward Morrill and Ward M. Hooper
Design Firm: Coleman, LiPuma, Segal & Morrill Inc., New York, NY
Client: Pfizer Inc., New York, NY

The logotype, with its transitional warm tones, symbolizes the product's effectiveness as a vaporizing rub.

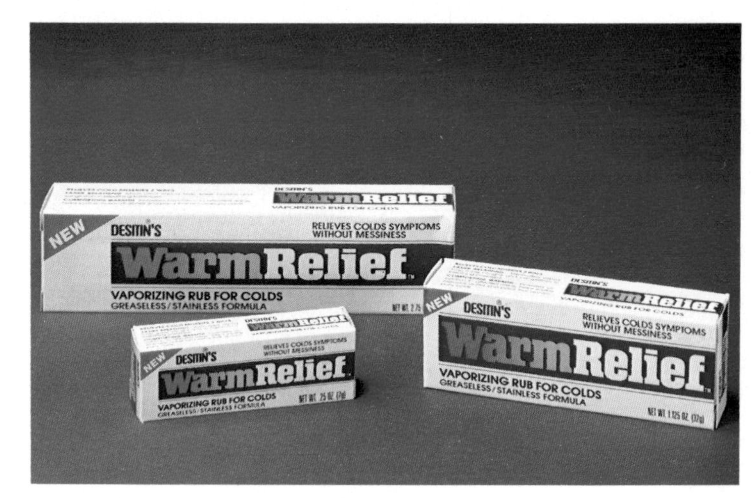

Product: Ticalma
Designer: Staff, Kelemata
Client: Kelemata, Torino, Italy

These Kelemata products are packaged in pale blue and white boxes. The effect is calming and soothing. Each tablet is blister packed for easy dispensing.

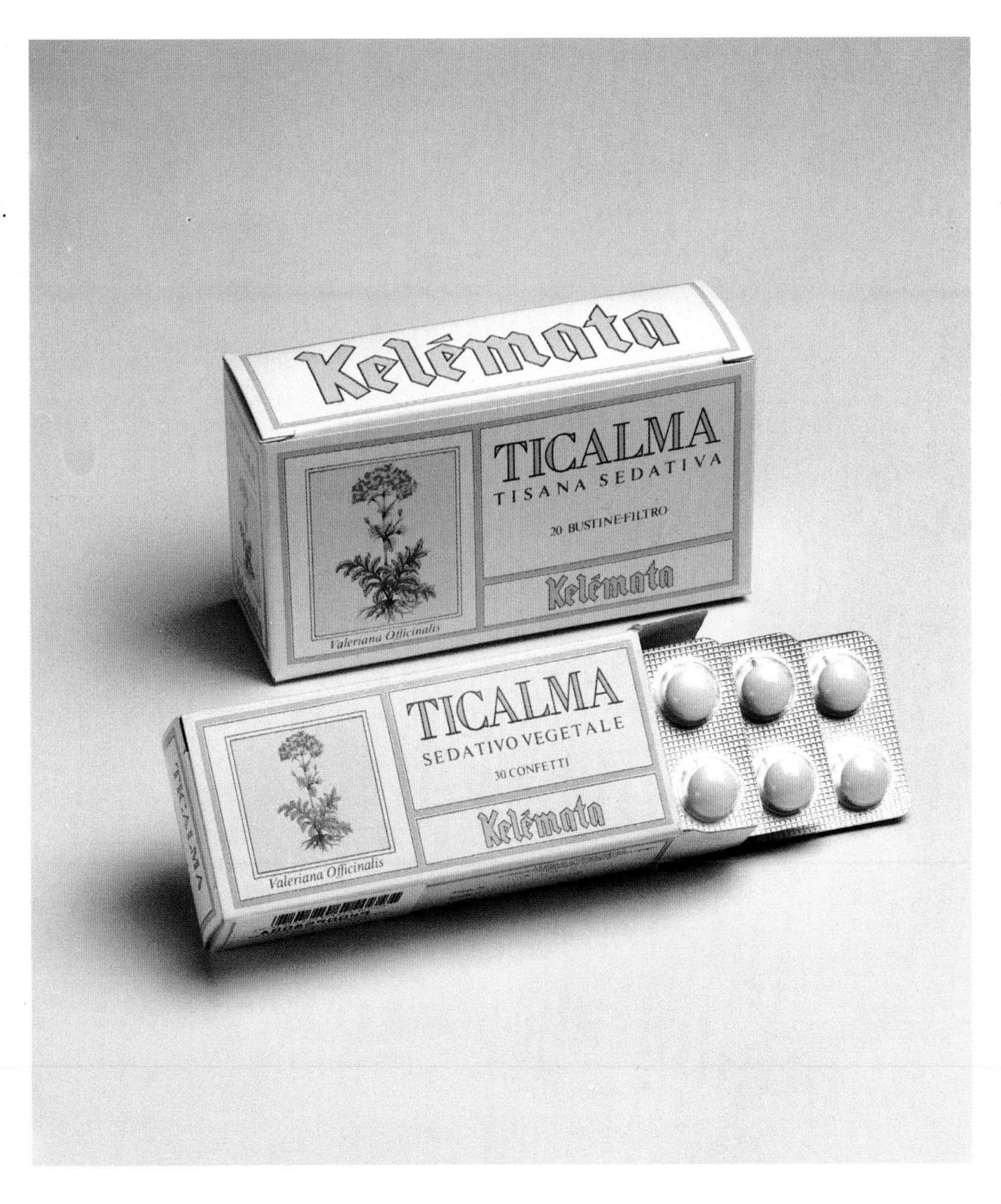

Product: Mediplus First Aid Products
Designer: Piero Gatti
Design Firm: Armando Testa S.p.A., Torino, Italy
Client: A.C.R.A.F., S.p.A., Cremona, Italy

Clearly illustrated first aid packages make it easy for consumers to choose what they need.

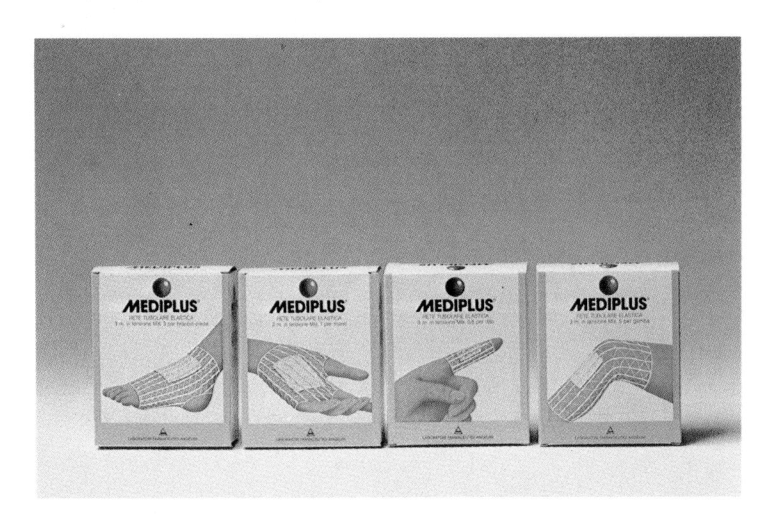

Product: Ben Gay Sports Gel
Designers: Edward Morrill and Ward M. Hooper
Design Firm: Coleman, LiPuma, Segal & Morrill Inc., New York, NY
Client: Pfizer Inc., New York, NY

A contemporary sports-oriented graphic image for a "before and after exercise" rub attracts the young, active consumer already familiar with the product.

Product: Baby Care Products
Designer: Mary Lewis
Design Firm: Lewis Moberly, London, England
Client: Tesco, Herts, England

The packaging for these baby care products uses an infant theme—pink and blue elephants and humorous antics.

Product: Iridina Blu Eye Drops
Designer: Staff designers, Montefarmaco S.p.A.
Design Firm: Montefarmaco S.p.A.
Client: Montefarmaco S.p.A., Italy

The box illustrates chamomile and hamamelis extracts, the key ingredients in Iridina Blu eye drops.

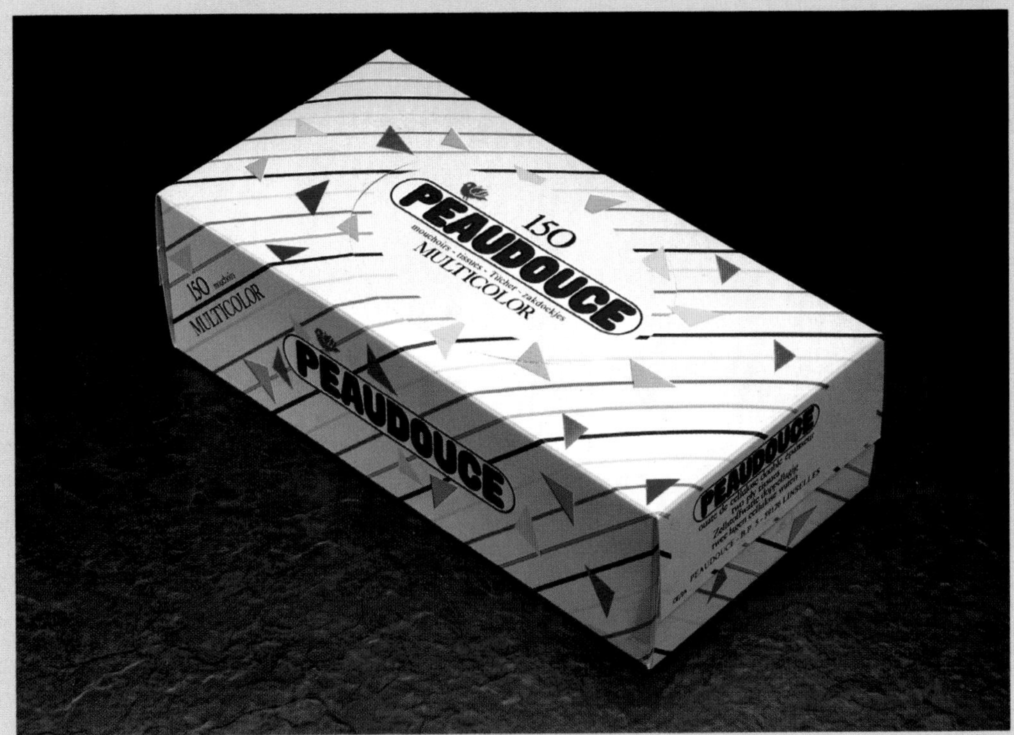

Product: Peaudouce
Designer: Paule Vezinhet and Pierre LeGonidec
Design Firm: GE2A, Lille, France
Client: Peaudouce B.S.F., Linselles, France

Peaudouce, meaning soft skin, is packaged in a multi-colored box.

Product: Feminine Hygiene
Designer: Silvano Lana
Design Firm: Armando Testa S.p.A., Torino, Italy
Client: Fater S.a.s., Pescara, Italy

Delicately-feminine illustrations enhance a line of feminine hygiene products. The bottle of liquid soap is designed for easy use.

Product: Alaxa Laxative
Designer: Piero Gatti
Design Firm: Armando Testa S.p.A., Torino, Italy
Client: A.C.R.A.F., S.p.A., Cremona, Italy

Graphics indicate that this product acts in the lower intestines, where needed. Electric blue evokes a feeling of comfort, relief, and rest.

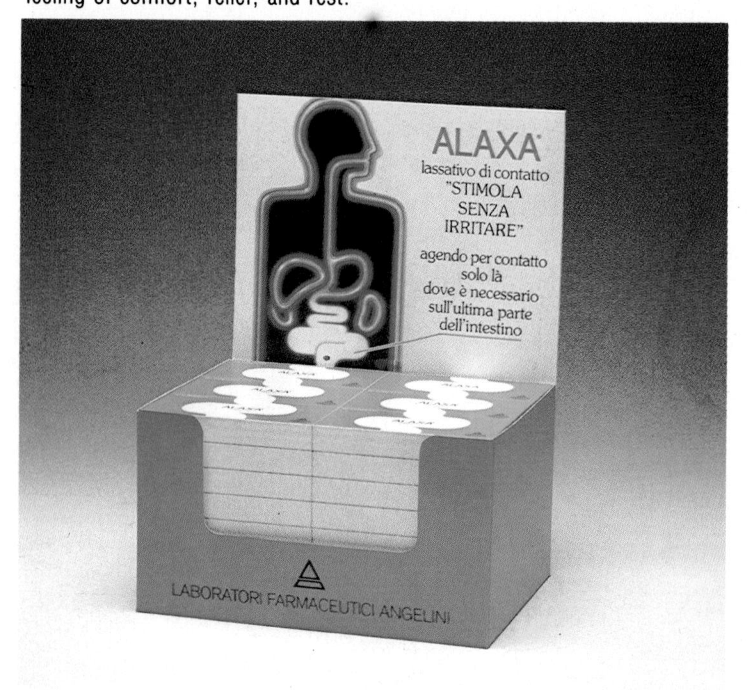

Product: Feminine Napkins
Designer: Giampiero Ferrari
Design Firm: Armando Testa S.p.A., Torino, Italy
Client: Fater S.a.s., Pescara, Italy

These beltless napkins are contained in an easy-to-carry package. An illustration of a grown woman's eyes distinguishes the package from a similarly shaped diaper package.

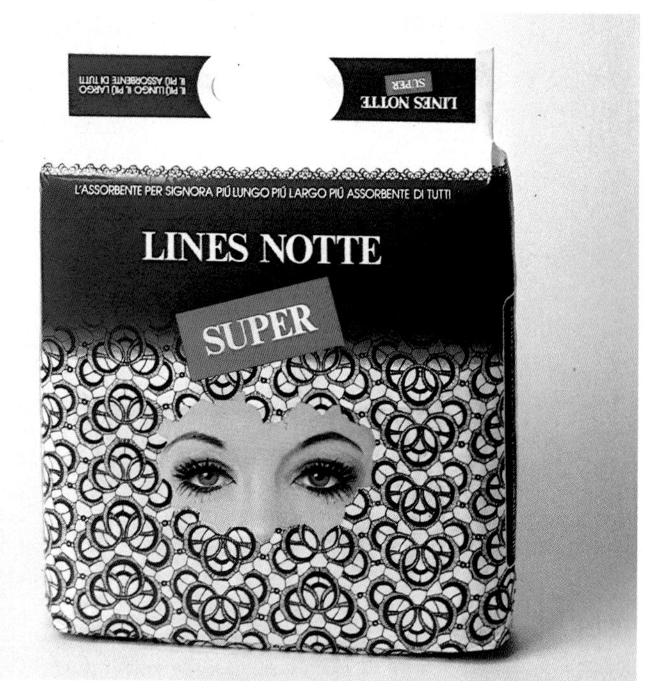

Product: Lacoste Toiletries for Men
Designer: Alain Carré Design SA.
Design Firm: Alain Carré Design SA., Paris, France
Client: Lacoste, France

The renowned alligator symbol promotes this Lacoste after-shave, giving it positive brand identification.

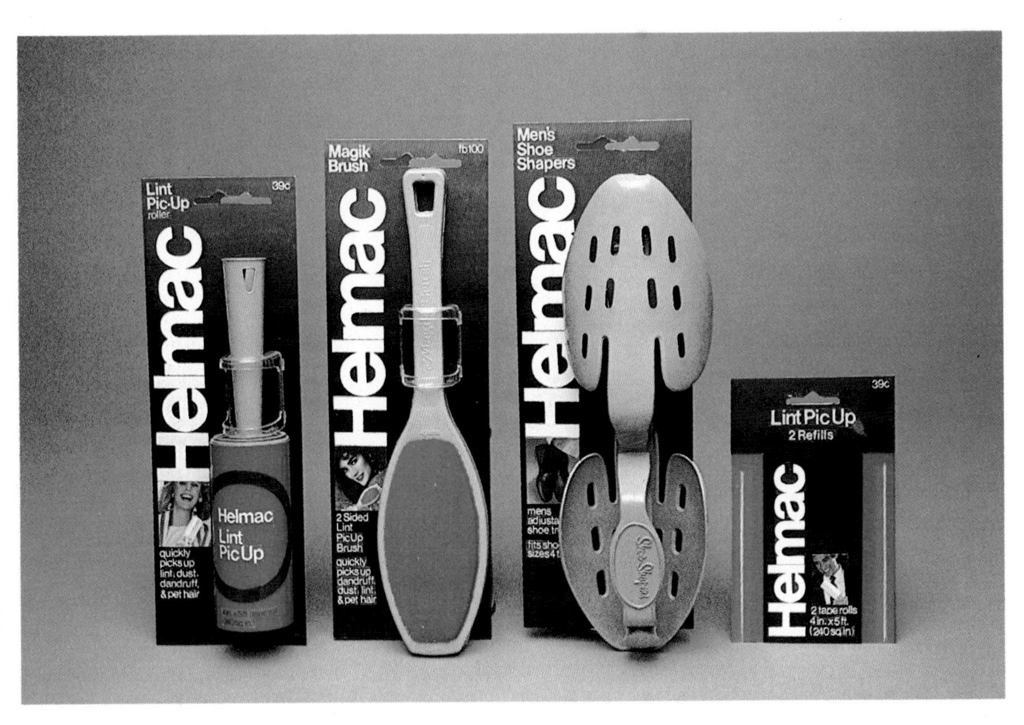

Product: Helmac Products
Designer: Jerry Dior
Design Firm: Gerstman & Meyers Inc., New York, NY
Client: Helmac Products Corporation, Warren, MI

Helmac redesigned their existing package to reflect uniformity, product identification, and better brand identification. The logotype was designed in white, with a new red background color.

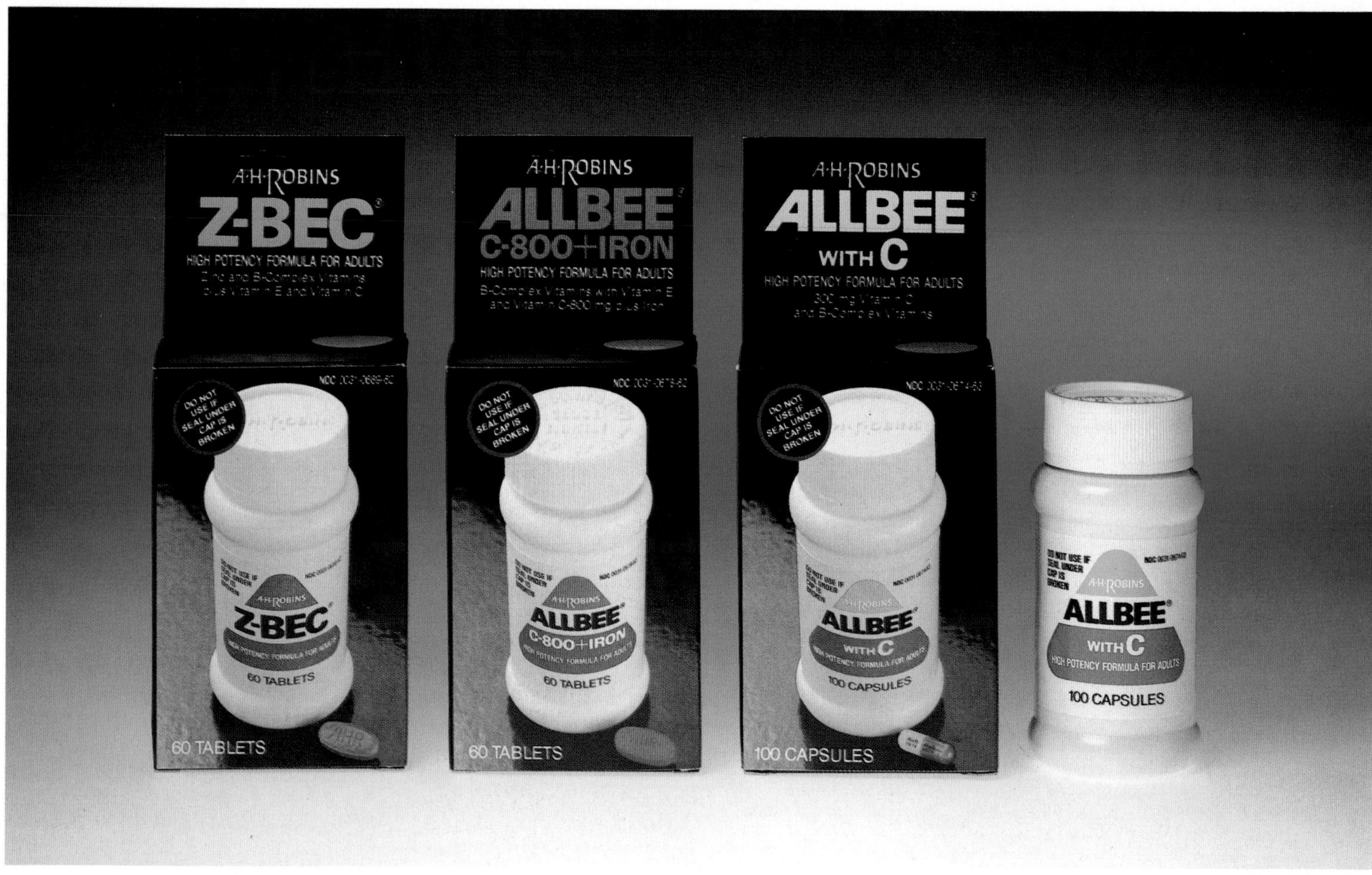

Product: Allbee/Z-Bec Vitamin Line
Designers: Owen W. Coleman and John Rutig
Design Firm: Coleman, LiPuma, Segal & Morrill Inc., New York, NY
Client: A.H. Robins Company, Richmond, VA
Award: 1984 Clio Award for packaging.

Designed as a vitamin container, this new boxboard structure has color coding against a black background to separate the individual vitamins.

Product: Capri Shampoo
Designer: Linda Voll
Design Firm: Murrie White Drummond & Lienhart Associates, Chicago, IL
Client: Northern Laboratories Inc., a Johnson Wax Company

To help move Capri beyond its traditional discount outlets into mainline chain stores, a new, higher quality brand image was created to offset the previous generic-like image.

Product: Dietary Supplements
Designer: Invernizzi Ottorino
Design Firm: Invernizzi Ottorino, Milano, Italy
Client: Serono OTC S.p.A., Milano, Italy

The package for this product, which is sold mainly in pharmacies, is highly visible on a store shelf. Each tablet is individually wrapped inside a box that clearly illustrates the name of the product.

Product: Toni Silkwave Perms
Designers: Edward Morrill and Ward M. Hooper
Design Firm: Coleman, LiPuma, Segal & Morrill Inc., New York, NY
Client: The Gillette Company, Boston, MA

A bold brushstroke promotes a well-established product name set against a series of colorful transition-tone backgrounds to differentiate the types of perms.

Product: Get Set Hair Conditioners
Designer: Edward Rebek
Design Firm: John Racila Associates, Oak Brook, IL
Client: Alberto Culver Company, Melrose Park, IL

A bold logotype, label photography, and color-coded graphics convey the different product formulas and create a strong shelf presence.

Product: Vitalis
Designers: Richard Gerstman, Juan Concepcion, and Karen Corell
Design Firm: Gerstman & Meyers Inc., New York, NY
Client: Bristol-Myers Products, New York, NY

These Vitalis products were designed to draw the usual loyal audience to a new presentation of the line. Attention was paid mainly to coloring strategy, trying to create a contemporary look while not losing too much of the old Vitalis coloring that helps identify the line.

Product: Airspray
Designer: Staff, Airspray International
Design Firm: Airspray International, New York, NY
Client: Airspray International, New York, NY

This new development is called the "Airspray System," an air-pressurized spray system that sprays like an aerosol, but operates without gas.

Product: Freeze-Dried Products for Intraveneous Injections
Designer: Groupe Design MBD
Design Firm: Groupe Design MBD, Paris, France
Client: CNTS, Paris, France

This simple, professional design is for freeze-dried concentrated products for intraveneous injections. The different colors on the boxes indicate specific doses.

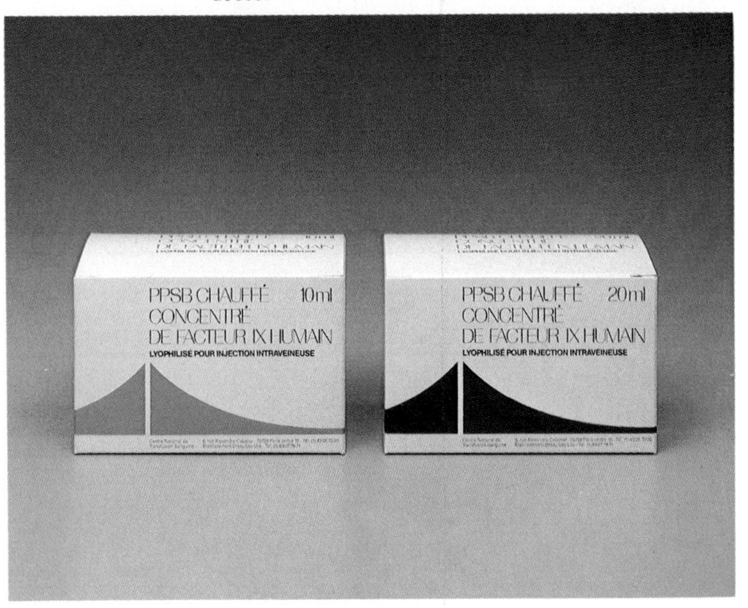

Product: Wet 'N Wild Bath Line
Art Director: Thomas D'Addario
Designers: Andrea Brooks and Linda Chow
Design Firm: The Creative Source Inc., New York, NY
Client: Pavion Ltd.

This new bath line consists of a splash cologne and a perfumed talc. The essence is floral, the coloring is pink, and the appeal is youthful.

Product: Protein 29
Designers: Owen W. Coleman and John Rutig
Design Firm: Coleman, LiPuma, Segal & Morrill Inc., New York, NY
Client: The Mennen Company, Morristown, NJ

An upgraded package design was created for all Protein 29 products, retaining some of the existing equities while giving it a more contemporary image.

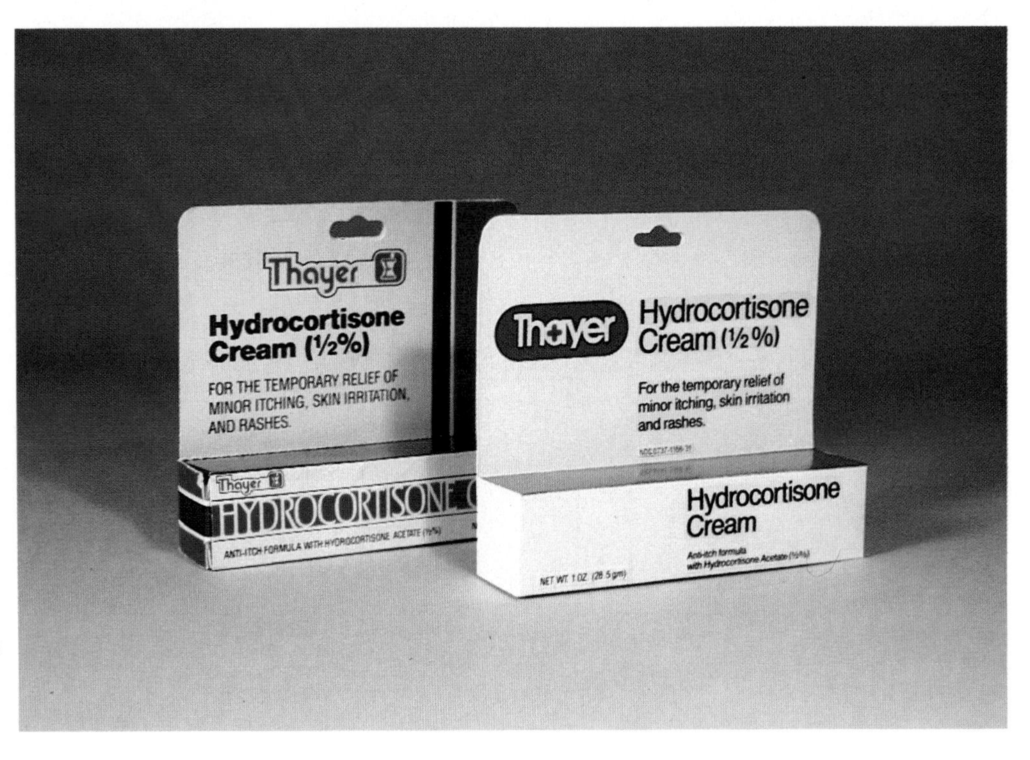

Product: Thayer Pharmacy
Designer: Selame Design
Design Firm: Selame Design, Newton, MA
Client: Thayer Pharmacy, Stoughton, MA

To implement the new Thayer Pharmacy corporate identity, all packaging was updated to reflect the new Thayer look. This package states clearly what the product is and what the package contains.

Product: Cosmetics, Toiletries
Designer: Mary Lewis
Design Firm: Lewis Moberly, London, England
Client: The Boots Company Ltd., Nottingham, England

Aimed at a teenage audience, these products are packaged in containers that suggest a mature theme. The container for Kohls is a cigar tin.

Product: Bath Salts
Designer: Mary Lewis
Design Firm: Lewis Moberly, London, England
Client: Tesco, Herts, England
Award: 1986 Clio nomination, International Packaging

Here, bath salts are packaged in a box with soothing, relaxing illustrations. The mood created is comparable to that experienced while relaxing in a warm bath.

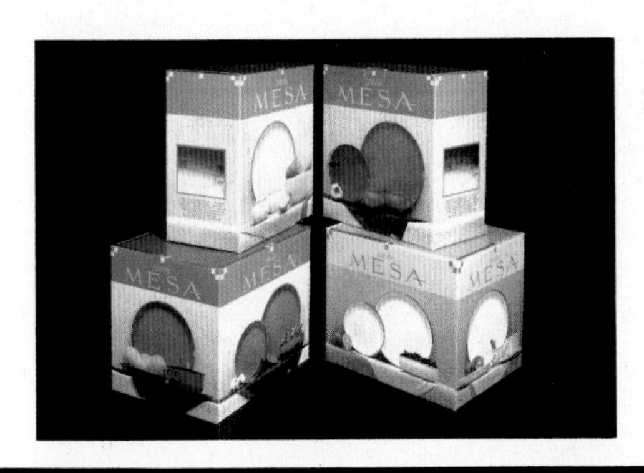

HOUSEWARES

The designers of packaging for housewares, a category of products that includes kitchenware, tabletop products, and appliances, are currently facing an important new challenge. Until recently, these products were stored and sometimes displayed in plain shipping cartons lacking outer graphics. But now, because department stores and mass merchandisers have dramatically reduced the number of sales clerks behind their counters, packaging must become a "silent salesman" that describes and promotes the product.

This requirement has led designers to the most basic grid design principles and to Eurostyle graphics, both of which allow large amounts of information to be presented orderly and clearly.

Masses of type and large product photographs are particulary effective on this new outer packaging. Generally, the amount of type on the package depends on the complexity of the product—more copy is usually needed to explain and promote appliances and other electrical or mechanical products. Dishes and other tableware, which require little explanation and appeal to the consumer's sense of style, depend more on strong product photographs to sell the product.

As can be seen from the examples in this chapter, the design industry is not only meeting the challenge of designing effective outer packaging but is creating truly exceptional designs. Among the most outstanding now displayed on retail floors are the package designs for the products of Dansk International Designs, Ltd., a firm that is careful to preserve its reputation for quality (pp. 146, 156, 157, 170, 171). Such excellent design undoubtedly plays a key role in attracting the consumer and, ultimately, in making the sale.

Exceptional package graphics and attractive product photographs promote the various cookware sets manufactured by Enterprise Aluminum Company. Here, the outer package features the name Vivre in stylized type printed in a striking red. Large photographs show off three cookware pieces while a smaller photograph displays the entire set. This package undoubtedly captures the consumer's attention and promotes the product's sale.

The packaging for Copco bowls allows the product to protrude, enabling the consumer to see its elegant design and feel its shape and texture. This technique effectively draws consumer interest and results in sales.

Product: Dansk "Neptune"
Designers: Glenn Groglio and Donald W. Moruzzi
Design Firm: Dansk International Design Ltd., New York, NY
Client: Dansk, Mount Kisco, NY

This pattern of Dansk glassware is symbolized by the god of the sea, Neptune. It is packaged in a blue box with the logo and Dansk name on the front.

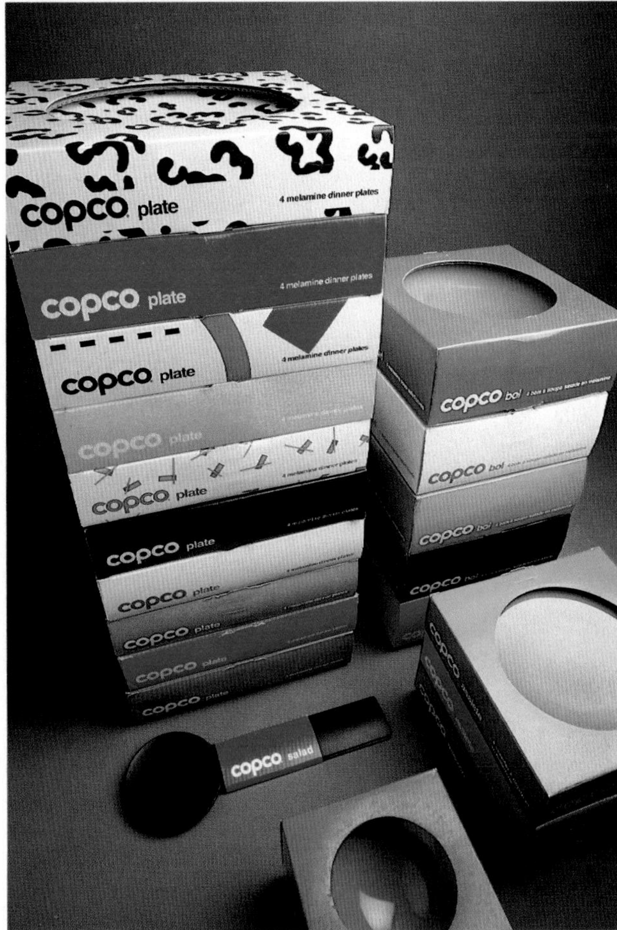

Product: Copco Melamine
Designers: Davin Stowell, A. Breckenfeld, T. Dair, B. Markee, and S. Russak
Design Firm: Smart Design Inc., New York, NY
Client: Copco, New York, NY

This packaging is meant to be an extension of the product. There is integration of form and color, and the inside comes out to show the contents.

SANYO

Café San V.I.P.™

10 Cup Electronic Coffee Maker with
Built-in Automatic Coffee Mill

Café San V.I.P.™ is easy to program.

9:46 Digital Electronic 24 hour clock

6:42 Set the exact time you want your fresh brewed coffee to be **ready**

10 Tell *Café San V.I.P.™* how many cups you want from 2 to 10

5 r Set *Café San V.I.P.™* to make your coffee the way you prefer, regular or strong

MILL Use the coffee mill manually or let *Café San V.I.P.™* do it for you automatically.

- Built-in mill grinds the beans just before brewing the freshest pot of coffee possible.
- Exclusive design gently pre-wets the ground beans evenly over the entire area before brewing.
- After grinding brewing process begins automatically . . . just set it and forget it.
- Easy to keep clean.

Sanyo Electric Inc.
200 Riser Road
Little Ferry, NJ 07643
USA

HOUSEHOLD USE ONLY

SAC 825

Café San V.I.P.™

10 Cup Electronic Coffee Maker with
Built-in Automatic Coffee Mill

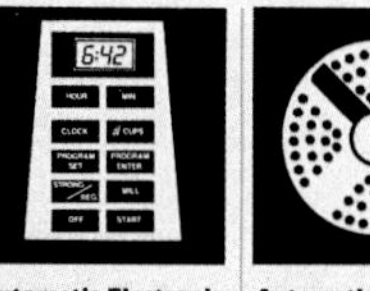

Automatic Electronic Controls
24 Hour digital clock allows you to program *Café San V.I.P.™* to have the exact number of cups ready when you want them.

Automatic Built-in Mill
Savor the aroma of freshly ground coffee beans while *Café San V.I.P.™* automatically grinds your favorite blend just before brewing to perfection.

Pre-wets Beans
Café San V.I.P.™ gently prewets the beans automatically after milling releasing the full flavor of the coffee beans.

Drip Guard
Allows you to take that "first cup" of coffee before the whole pot is finished . . . no more messy spills.

Two Hour Warmer
Our specially designed warmer keeps your coffee at the perfect serving temperature for two hours . . . then automatically turns off.

Self Storing Water Tank
The water reservoir removes for easy filling at the sink.

Permanent Filter
The exclusive dual filtration system eliminates the need for costly paper filters. Easy to clean, top rack dishwasher safe.

Automatic Cord Rewind
Allows for built-in cord storage to keep your counter neat.

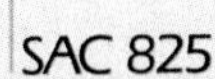

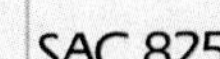

SAC 825

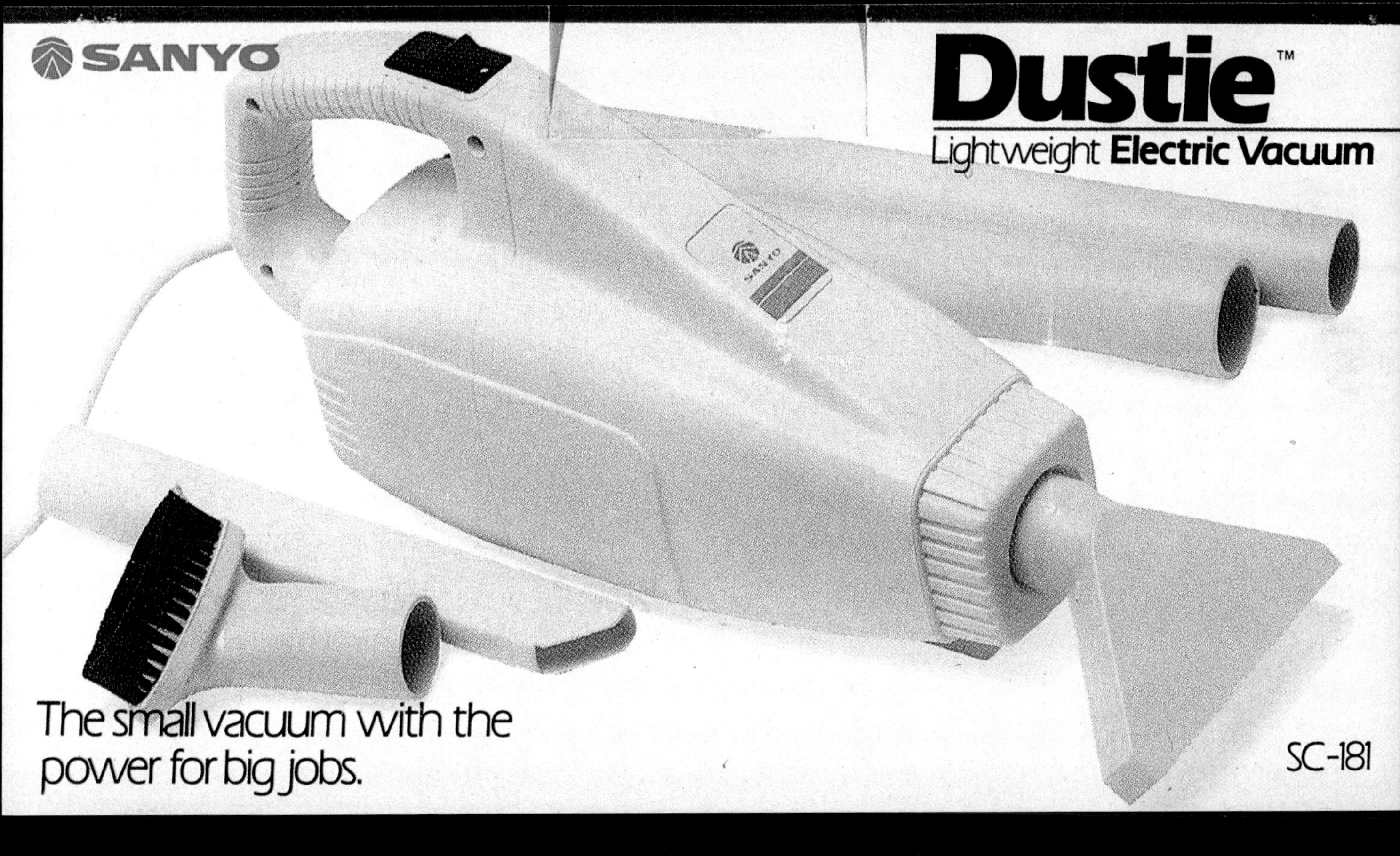

SANYO
Dustie™
Lightweight Electric Vacuum
The small vacuum with the power for big jobs.
SC-181

SANYO
Café San V.I.P.™
10 Cup Electronic Coffee Maker with
Built-in Automatic Coffee Mill
Built-in Automatic Coffee Mill
Dual Filtration System
Dripguard Valve
10 Cup Heat Resistant Carafe
24 Hour Programmable Electronic Control
2 Hour Warmer
10 Cup Water Reservoir
Cord Rewind Mechanism
Coffee Measuring Spoon
SAC 825

Product: Sanyo Packaging System
Designers: Stephen Jobe, Lauren Giber, David Bragin, and Bruce Wasserman
Design Firm: Bruce Wasserman & Associates, New York, NY
Client: Sanyo Electric, Little Ferry, NJ

This clean packaging system quickly identifies the specific category for Sanyo. All consumer information is clearly printed on the package.

Product: Barazzoni Fratelli Pots and Pans
Designer: Ennio Lucini
Design Firm: Studio Elle, Milano, Italy
Client: Barazzoni Fratelli, Italy

The Gran Cucina package is simple yet clear. The package for Glitter pots and pans is easy to carry and very attractive. The Tummy collection is securely packed in a smaller cardboard container.

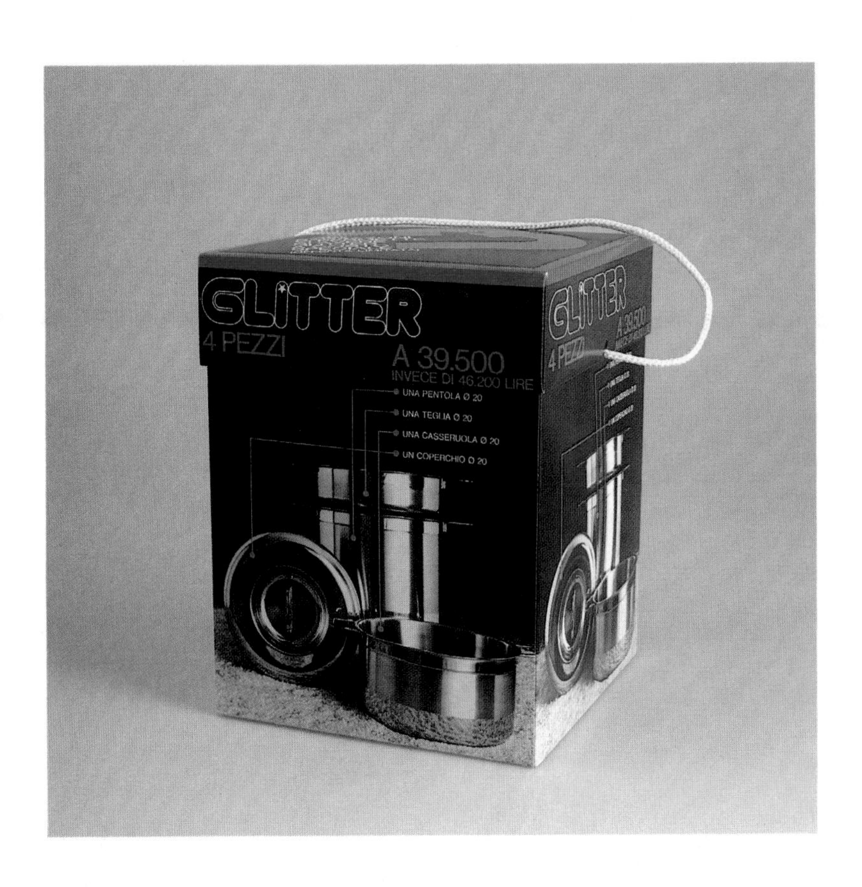

Product: Crema e Gusto Coffee Pot
Designer: Otello Fraternale
Design Firm: Armando Testa S.p.A., Torino, Italy
Client: Lavazza S.p.A., Torino, Italy

An illustration of the product and a cup of coffee set against a backdrop of red and blue subconsciously express the idea of heat. The steam rising from the coffee cup adds to this illusion.

Product: Veneziano Drinking Glasses
Designer: Staff designer Rocco Bormioli
Client: Rocco Bormioli, Italy

Veneziano red wine stem glasses, which are exported abroad, are packaged in a simple but elegant box. The Loto set is packaged in a colorful box, suggesting that it is appropriate for soft drinks and juices. The use of computer graphics for the Frigoverre containers indicates the more technical purpose of the product.

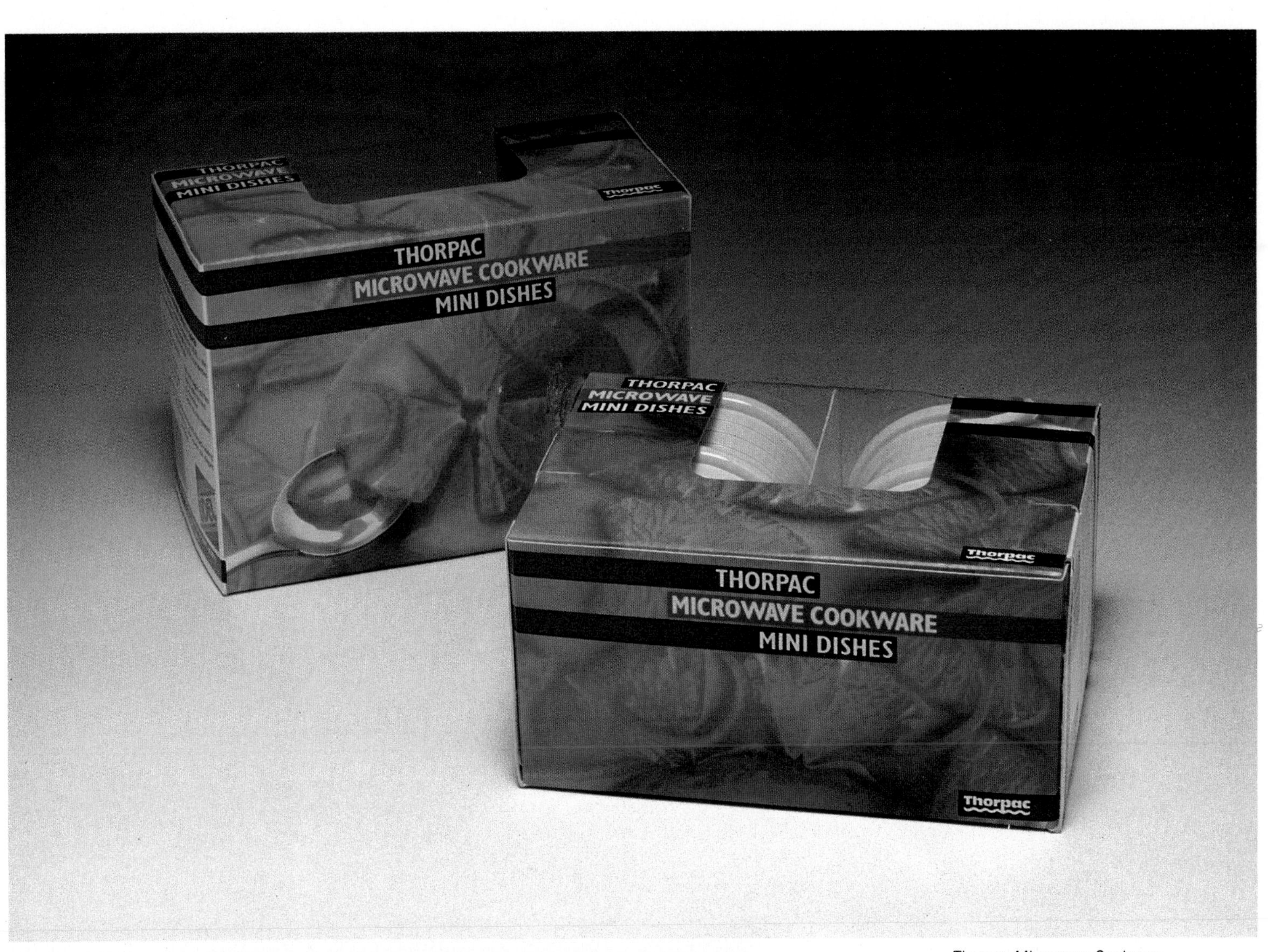

Product: Thorpac Microwave Cookware
Designers: Mervyn Kurlansky and Herman Lelie
Design Firm: Pentagram, London, England
Client: Thorpac

Banded typography and the composition of the different photographs allows for continuity when these microwave cookware packages are displayed side by side.

Product: Nescafé Coffee Container
Designer: Joël Desgrippe
Design Firm: Desgrippe, Beauchant, Gobé, Cato, Paris, France
Client: Sopad-Nestlé
Awards: Oscar de l'Imballage, Janus de l'Institut Francais du Design Industriel

Consistent with the Nescafé look, this coffee container features rounded edges. The container is refillable and unique to its product category.

Product:	Avent Infant Feeding System
Designers:	Mervyn Kurlansky and Herman Lelie
Design Firm:	Pentagram, London, England
Client:	Cannon Babysafe Ltd.

The brightly colored Avent logo stands out on a display shelf. This line, now consisting of nine items, was recently expanded.

Product: Dansk Dinnerware Sleeve and Foam Packaging
Art Director: Glenn Groglio
Designers: Todd Ressler and Danisha Devor
Design Firm: Dansk International Designs Ltd., New York, NY
Client: Dansk International Designs Ltd., New York, NY

This package is particularly useful for storing and shipping Dansk dinnerware.

Product: Open Country Camp Cookware
Designer: Edward Rebek
Design Firm: John Racila Associates, Oak Brook, IL
Client: Mirro Aluminum Company, Manitowoc, WI

Inviting photography capitalizes on the allure of the outdoors, while a bold headline quickly communicates the use of the product.

Product: Dansk Marching Band
Designers: Staff designers Glenn Groglio and Carol Haley
Client: Dansk International Designs Ltd., New York, NY

An elongated, rectangular box shows the product through a cellophane window. The marching band is brought to life by a backdrop of a town, through which the band seems to be moving.

Product: Nubril Floor Cleanser
Designer: P.G.J.
Design Firm: P.G.J., Paris, France
Client: Ciba-Geigy, France

This no-rinse floor cleanser is packaged in a convenient plastic bottle that is easy to carry and to handle.

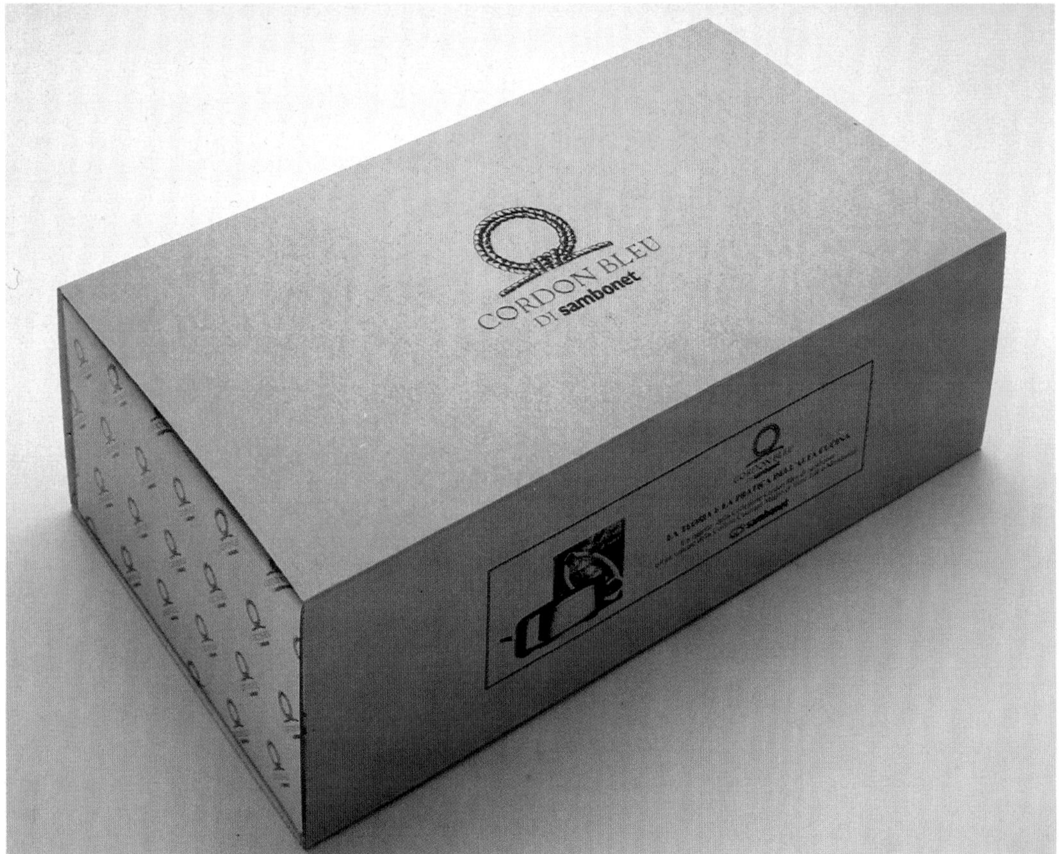

Product: Cordon Bleu Sambonet Flatware
Designer: Roberto Sambonet
Client: Sambonet, Italy

Sambonet manufactures different lines of flatware. The new packages are designed with low cost and high style in mind.

Product: Baker's Secret
Designer: Linda Voll
Design Firm: Murrie White Drummond & Lienhart Associates, Chicago, IL
Client: Ekco Housewares Inc.

The Baker's Secret redesign contemporizes the existing brand mark while protecting existing visual equities and adding strong appetite appeal with new photography.

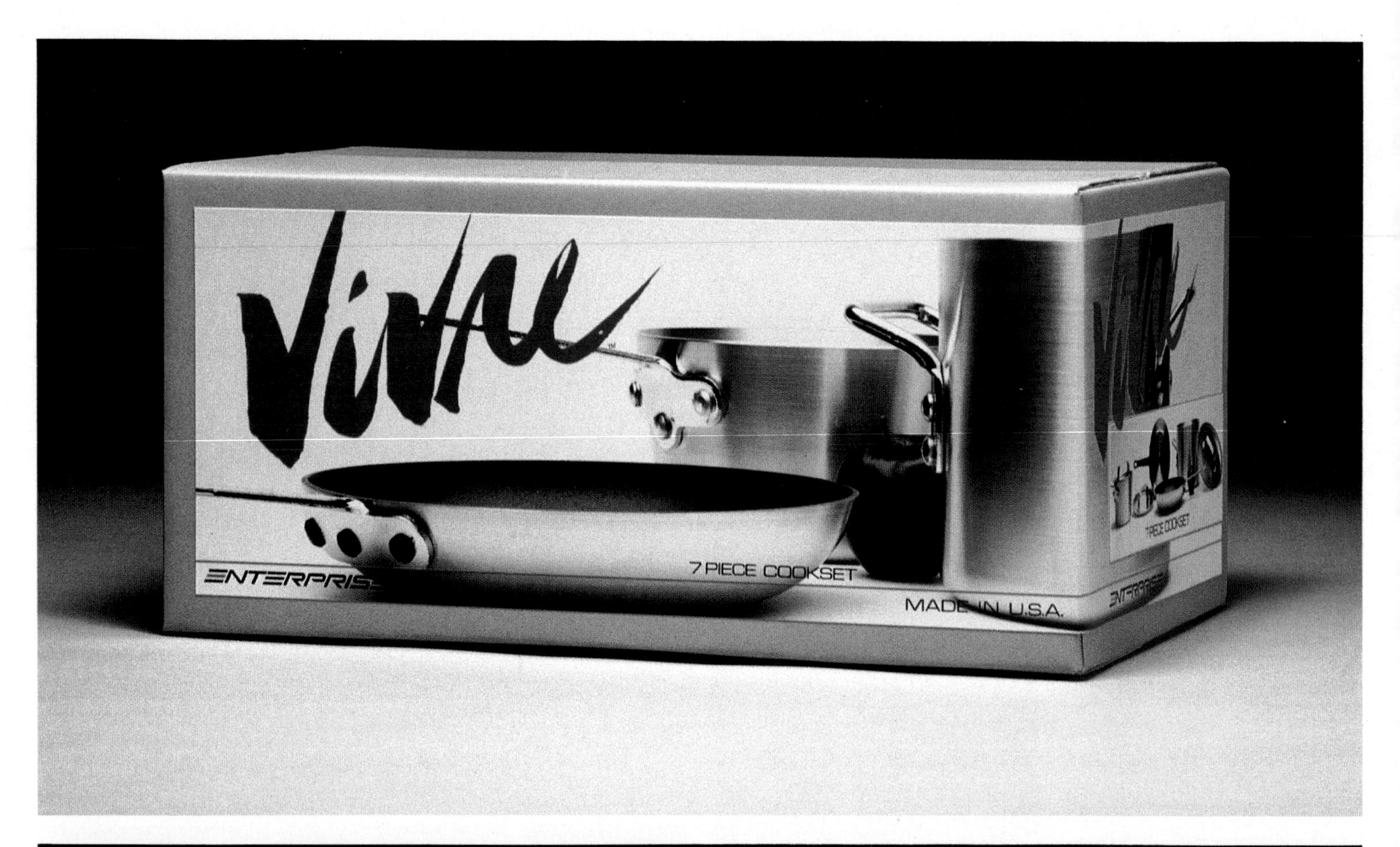

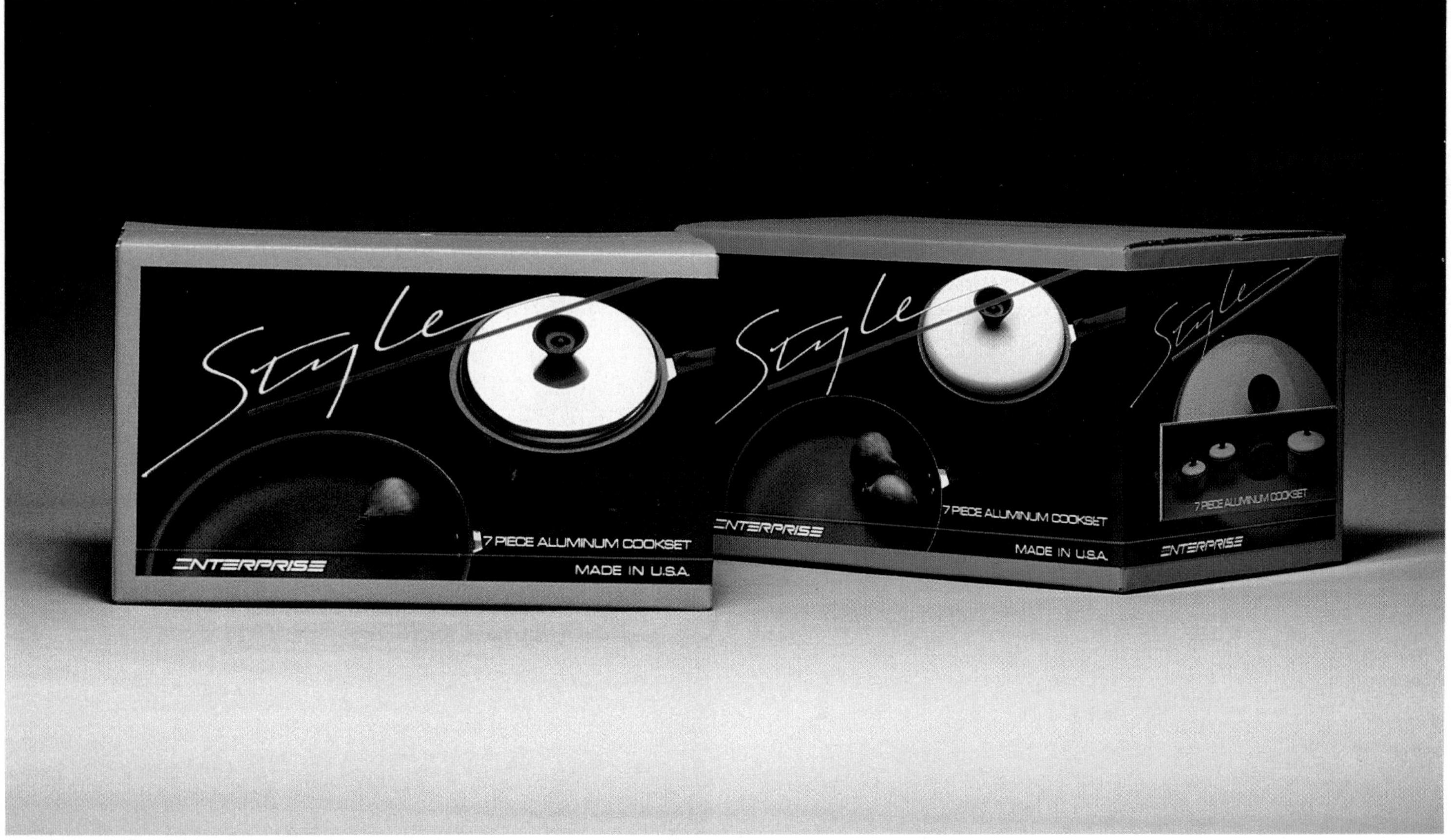

Product: Enterprise Aluminum Cookware
Designers: Laura Garza, Gus Methe, John Neher, and Edward Rebek
Design Firm: John Racila Associates, Oak Brook, IL
Client: Enterprise Aluminum Company, Macon, GA

This line of cookware was designed to give each cookset its own identity while maintaining a cohesive family look. Each package has a unique, chic appearance.

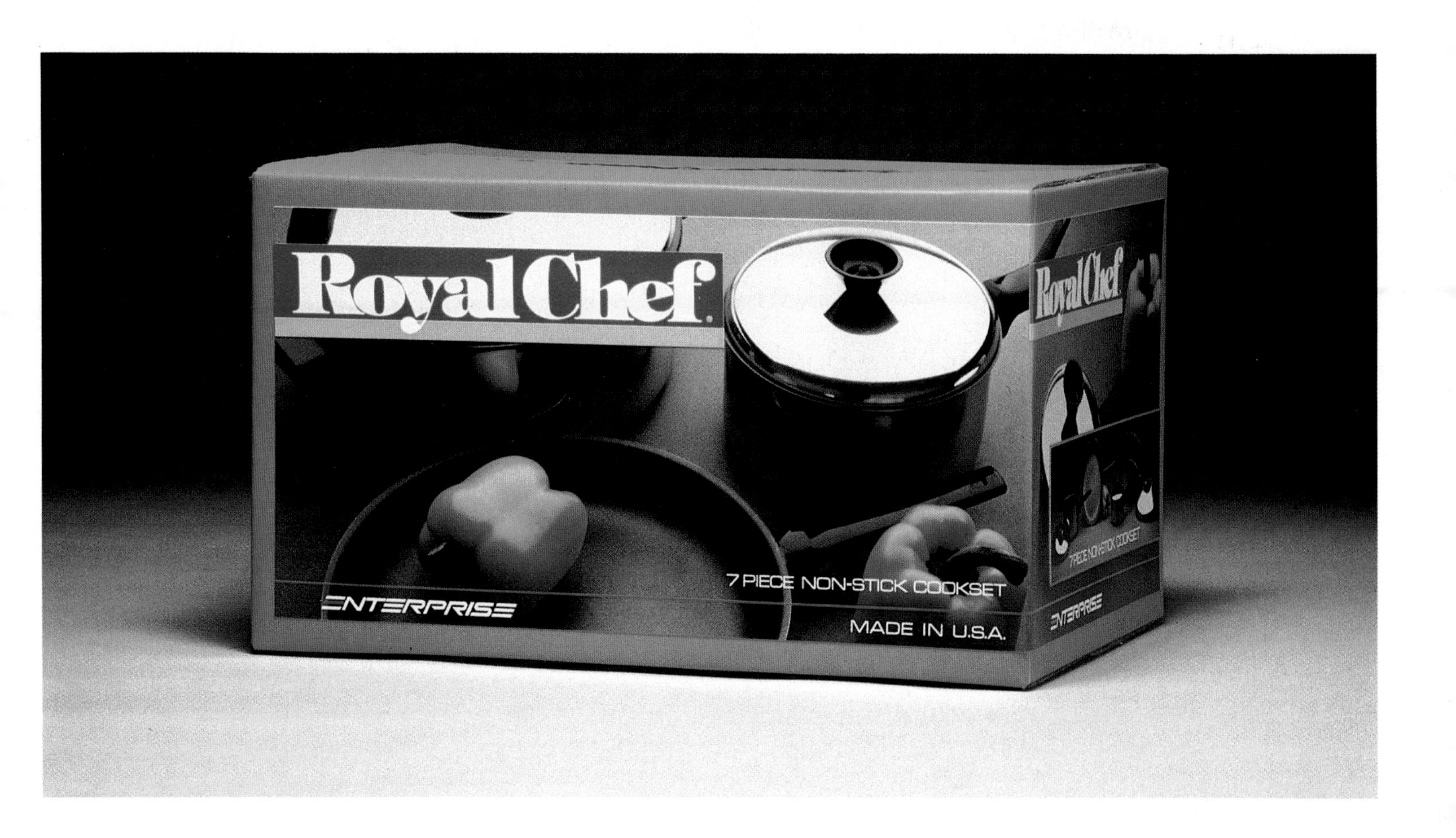

Product:	Necchi Domestic Appliances
Designers:	Giovanni Brunazzi and Giovanna Ceste
Design Firm:	Image and Communication, Torino, Italy
Client:	Necchi, Pavia, Italy

These packages for small appliances were designed so that the brand name is spelled out when the boxes are displayed next to one another. Large still-life photographs of different foods appeal to the consumer.

Product: Shout
Designer: Jeff White
Design Firm: Murrie White Drummond & Lienhart Associates, Chicago, IL
Client: S.C. Johnson & Son, Inc.

The original package was redesigned, but the new design creates the same dynamic appeal for the product that the original one had.

Product: Kenwood Appliances
Designers: Mervyn Kurlansky and Herman Lelie
Design Firm: Pentagram, London, England
Client: Kenwood

A uniform package design categorizes each separate appliance in the Kenwood line.

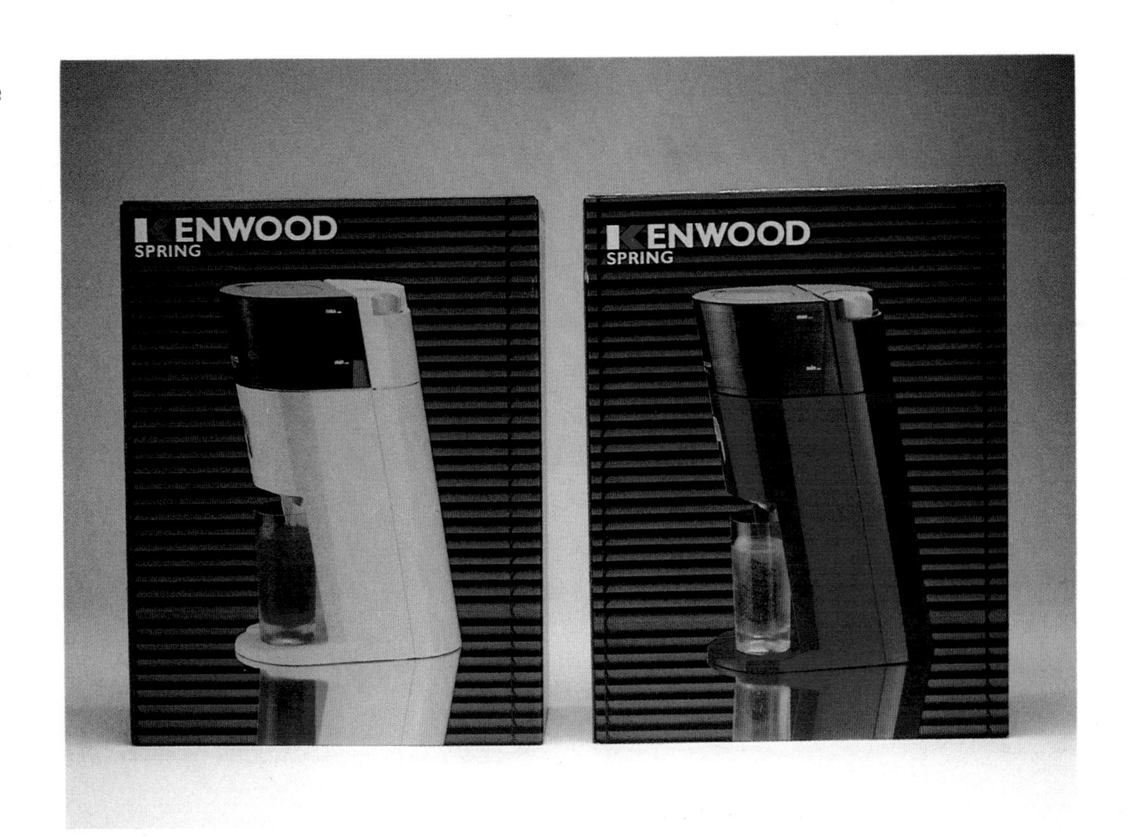

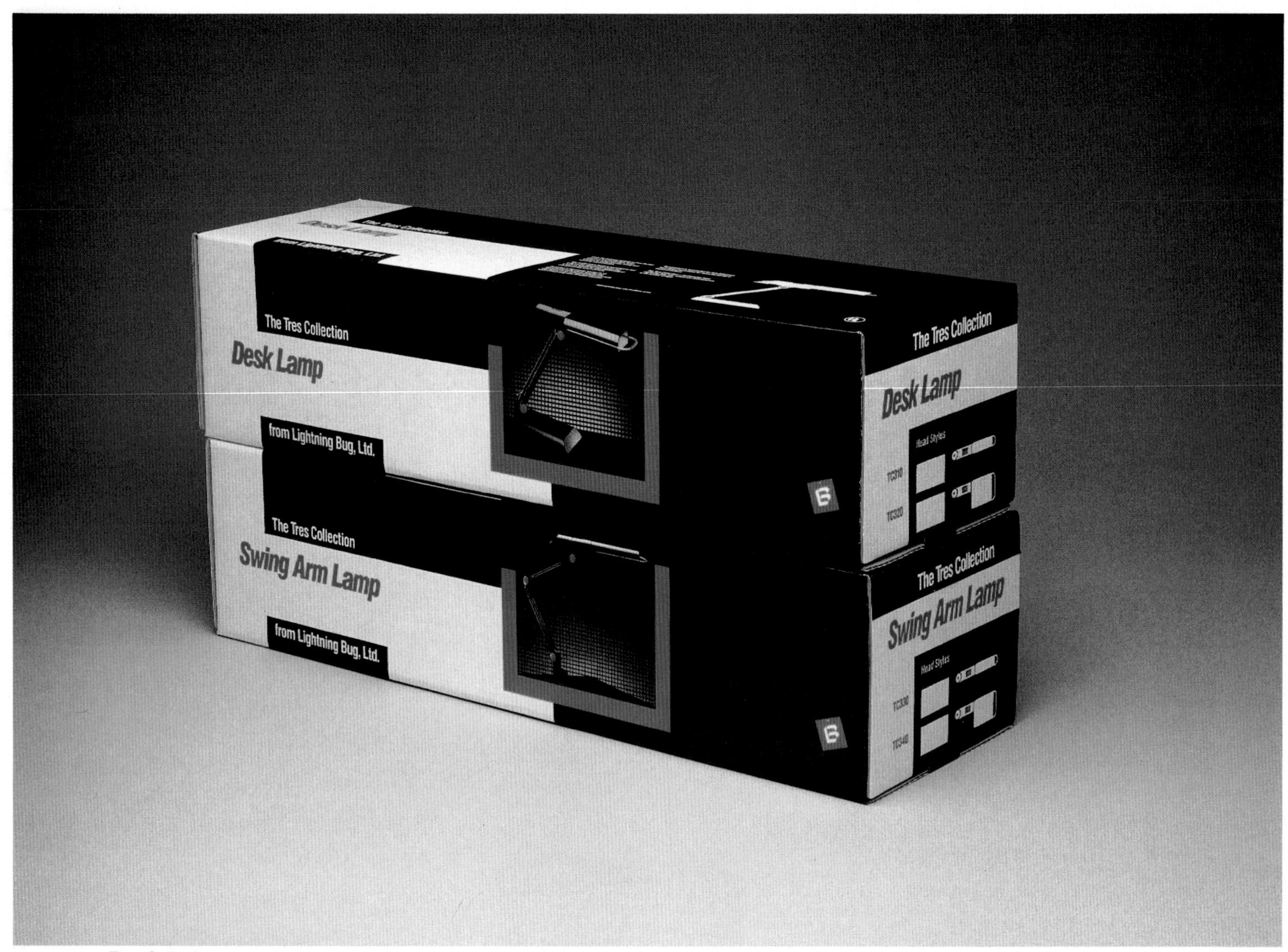

Product: Tres Collection
Designers: Allen Porter and Lisa Ellert
Design Firm: Porter/Matjasich & Associates, Chicago, IL
Client: Lightning Bug Ltd., Hazel Crest, IL

With its high-tech look and sound structure, this package is a successful display unit as well as an effective shipping box for the manufacturer.

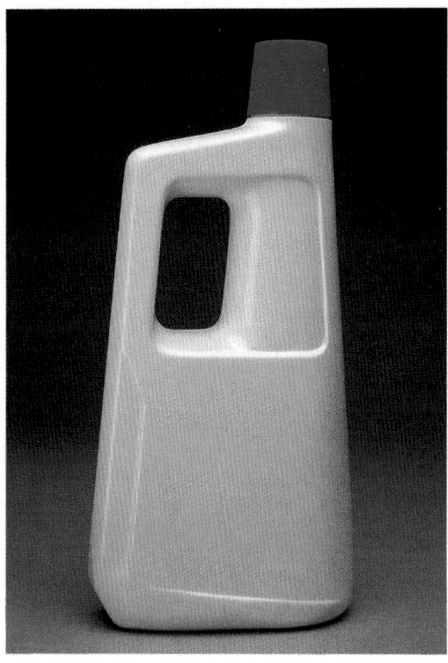

Product: Dynamo and Ajax
Designer: Ronald Peterson
Design Firm: Peterson Blyth Associates Inc., New York, NY
Client: Colgate

The packaging for both Dynamo and Ajax went through structural changes when designers created these easy-to-use containers. The strong side-grip handles reflect a hardworking image for both products.

Product: Vizir
Designers: Francois du Villard and Christophe Blin
Design Firm: Lonsdale Design & Communication, Paris, France
Client: Procter & Gamble, France

The cap on the Vizir bottle can be used to measure the product. It also fits onto the central piece of a standard washing machine.

Product: Solitaire Cotton Press
Designer: P.G.J.
Design Firm: P.G.J., Paris, France
Client: Solitaire, France

An easy-to-use spray for ironing cottons is packaged in a bright blue can that displays an illustration of an iron with wings. The illustration is intended to relate the product to the idea of quick, easy ironing.

Product: Lester Thread
Designer: Susanna Vellebona
Design Firm: Esseblu, Torino, Italy
Client: Cucirin Tre Steue, Italy

These spools of variously colored thread come packaged in a graphically designed cardboard box. The Lester logo, signifying current technology, is printed in white and also in pink.

Product: Carpet Cleanser
Designer: P.G.J.
Design Firm: P.G.J., Paris, France
Client: Ciba-Geigy, France

The label on this container of carpet cleanser is actually a photograph of a clean carpet.

Product: Performer 200
Designer: Groupe Design MBD
Design Firm: Groupe Design MBD, Paris, France
Client: Delta Dore, France

This portable thermostat is packaged in a high-tech fashion to indicate that it is an electrical appliance. The colored waves indicate different temperatures.

Product: Dishwashing Powder
Designer: Alain Carré
Design Firm: Alain Carré, Paris, France
Client: Solitaire, France

Solitaire is a small family-owned French company that manufactures household products. The company introduced Opalor dishwashing powder to the market, in competition with two large companies. To differentiate Opalor from similar products, Solitaire produced a plastic container. The contemporary package protects the product from moisture and features a built-in handle for additional convenience, a long neck for easy pouring, and colorful caps to differentiate the product. In addition, the containers are stackable.

Product: Salem China
Designer: Ed Szczyglinsky
Design Firm: Gerstman & Meyers Inc., New York, NY
Client: Salem China Company, Salem, OH

These festive Christmas products are manufactured by Salem China for sale under the "Noel" brandname. The main objective in designing these packages was to aim for point-of-purchase sales in department stores; therefore, a dynamic visual effect was necessary.

Product: Blankets
Designer: Luciano Lorenzi
Design Firm: G. & R. Associati, Milano, Italy
Client: Lanerossi, Vicenza, Italy

These blankets are packaged in convenient, ready-to-carry boxes. The larger box features a night sky, and the smaller one features a kite, which represents the "kite line" of Lanerossi products. The illustrations are soothing, much like a comfortable, warm bed is.

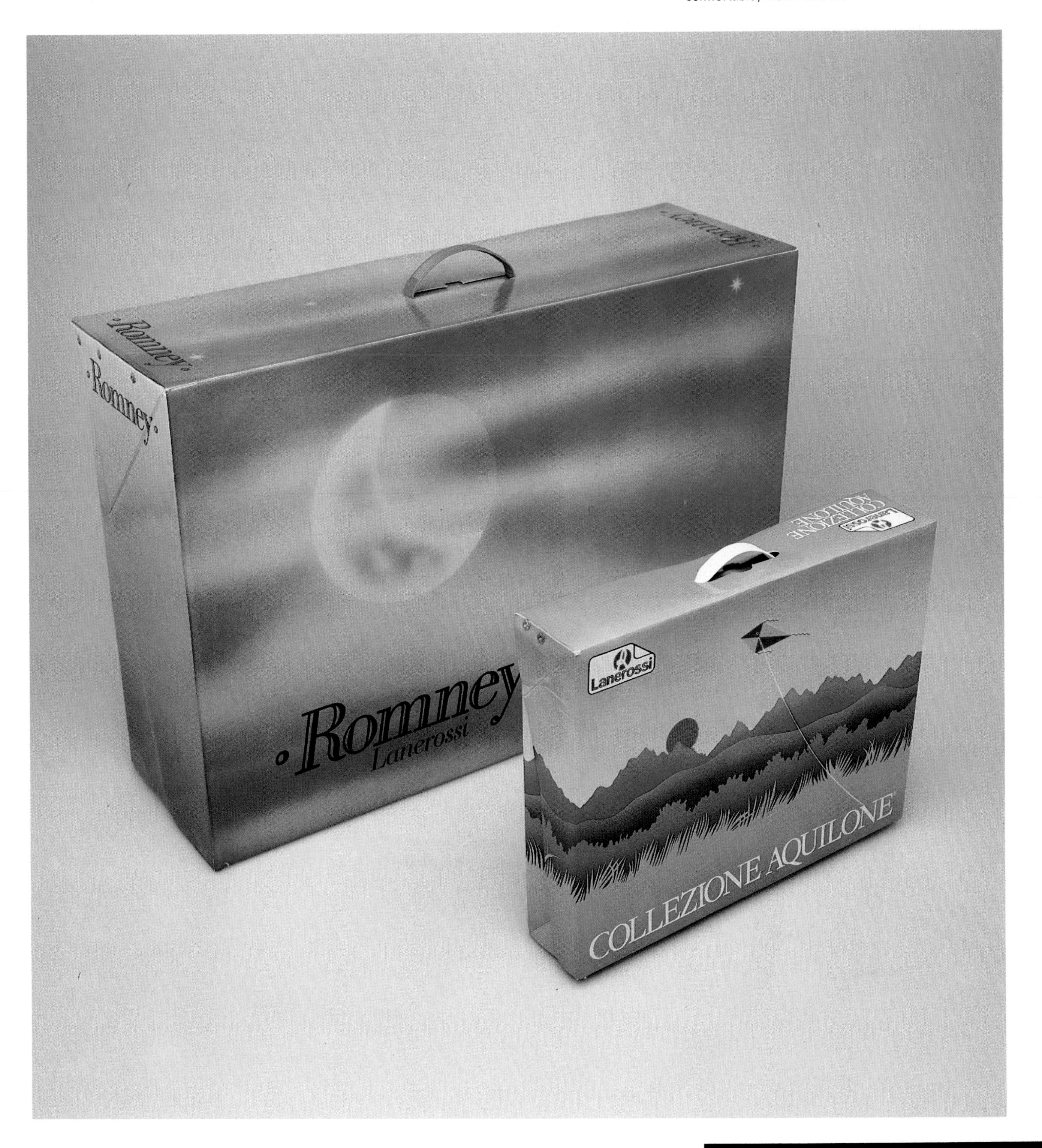

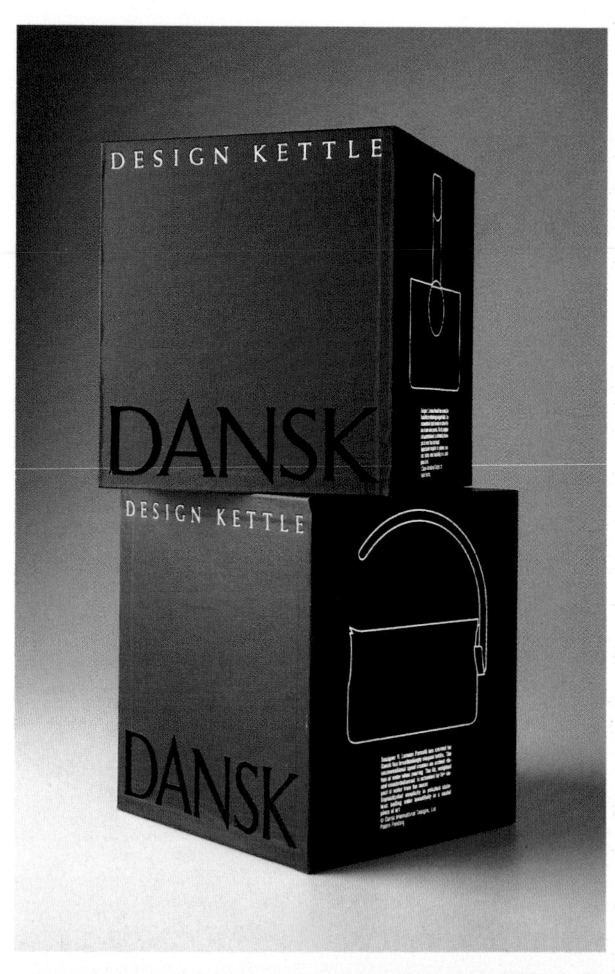

Product: Dansk Design Kettle
Designers: Staff designers Glenn Groglio and Donald W. Moruzzi
Client: Dansk International Designs Ltd., New York, NY

The dark Dansk name and the charcoal gray of the box are contrasted against a white illustration of the Dansk product.

Product: I Guzzini Glasses
Designer: Ennio Lucini
Design Firm: Studio Elle, Milano, Italy
Client: Guzzini Fratelli, Italy

These stackable glasses are packaged in a simple, yet contemporary box. Indentations of the corners of the boxes let the consumer know how many glasses are inside.

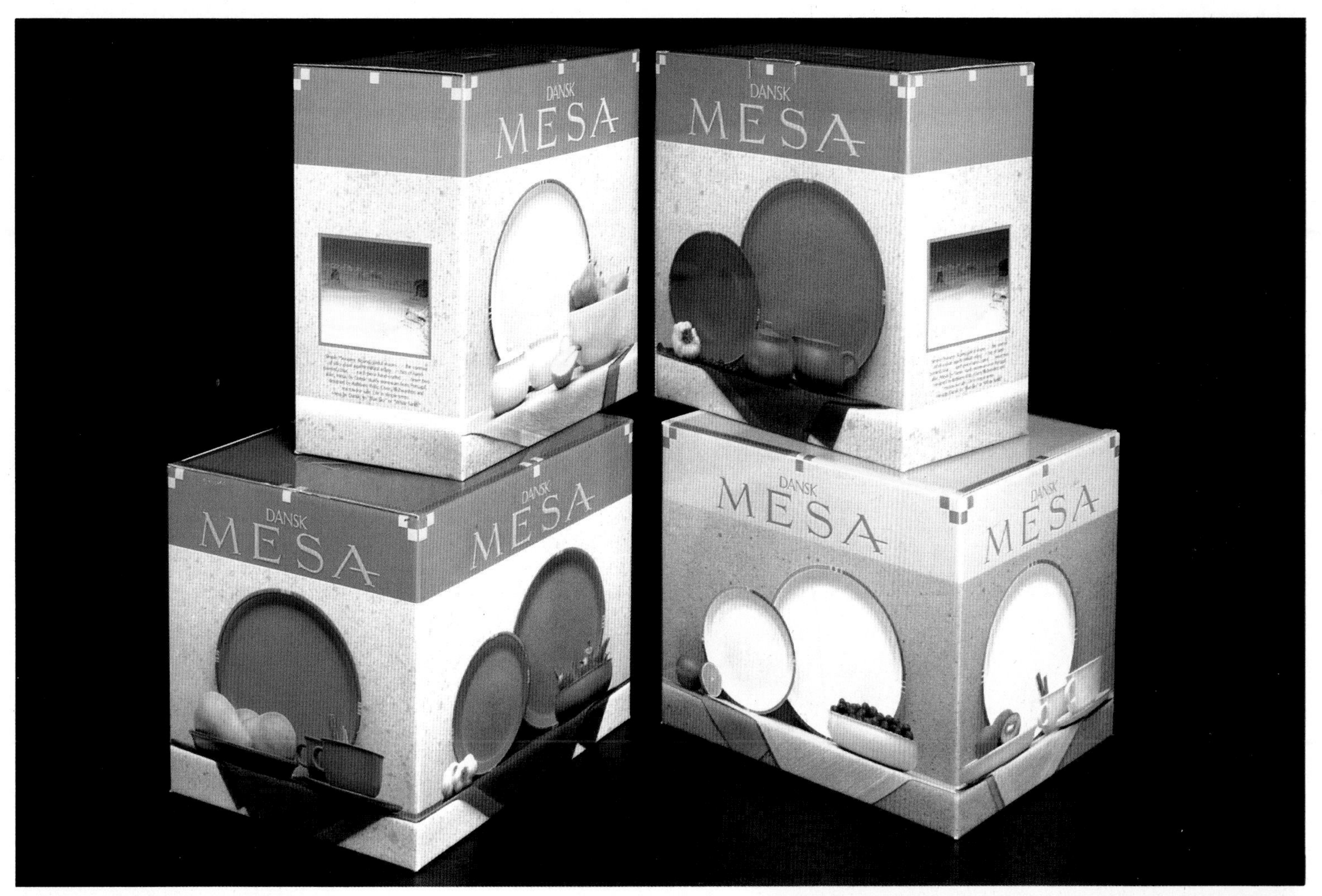

Product: Dansk Mesa Dinnerware
Designer: John Emery
Design Firm: Dansk International Designs, New York, NY
Client: Dansk International Designs, New York, NY

Uniformity of design was maintained in Dansk Mesa Dinnerware packages.

Product: Dansk Lightwoods
Designers: Staff designers Glenn Groglio and Donald W. Moruzzi
Client: Dansk International Designs Ltd., New York, NY

Different gradations of light gray were used on this packaging system for Dansk Lightwoods products.

Chapter 5

FINE GIFTS

WRAPPING PAPER
GIFT BOXES
DEPARTMENT STORES
STATIONERY

The high-margin products included in this chapter are geared for an upscale luxury market. Found in gift, stationery, and department stores, these items must be clothed in a package that conveys a sense of refinement and value. In fact, the packaging for this type of product is nearly as important as the product itself. It may even help the consumer justify a steep price.

This area of packaging affords designers the opportunity to do some of their most creative and elegant work. Among the most sophisticated designs now in this upscale marketplace is that of Lion stationery (p. 174), which, created by the design firm of Pentagram, features an elegant illustration and simple white type positioned against a rich blue background between subtle red rules. This is just one of the many outstanding designs that can be enjoyed along with the products they promote.

Classy and conservative, the packaging designed for Weather's Men's Clothier by the firm Semaphore Inc. features elegant, understated graphics. A tasteful illustration and simple type stand out against a rich wine-colored background.

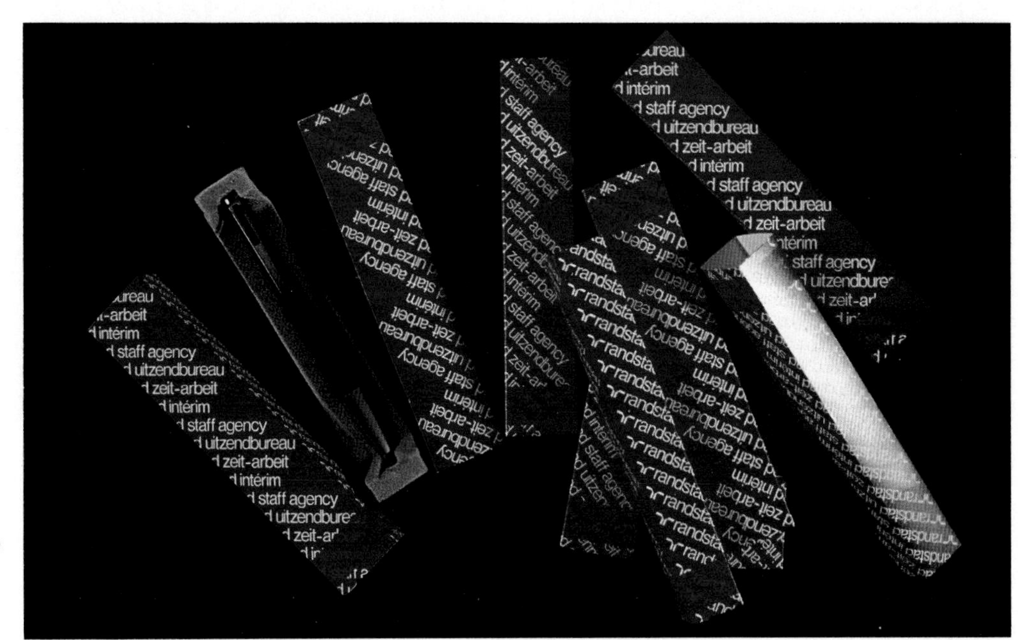

A contemporary typeface and the creative arrangement of words on a deep rose background give this pen giftbox an appealing, special look.

Product: Lion Stationery
Designers: Mervyn Kurlansky and Karen Blincoe
Illustrator: Chris Wormell
Design Firm: Pentagram, London, England
Client: DRG Stationery

In the redesign for the entire range of Lion stationery, a new symbol and a new system of color coding was adopted to identify the products within the line.

Product: Pen Giftbox
Designer: Zdena Srncova
Design Firm: Total Design, Amsterdam
Client: Randstad Uitzendbureau bv

This graphically appealing gift box holds a pen.

Product: Wineglass Giftbox
Designer: Zdena Srncova
Design Firm: Total Design, Amsterdam
Client: Randstad Uitzendbureau bv

Contemporary design for a wineglass giftbox serves to protect the product while being highly attractive.

Product: Pottery Barn Christmas Wrapping Paper
Designers: D. Stowell, T. Viemeister, T. Thomsen, J. Lonczak, and T. Dair
Design Firm: Smart Design Inc., New York, NY
Client: The Pottery Barn Inc., Elmsford, NY

Pottery Barn wrapping paper is graphically designed to capture seasonal cheer and warmth. It calls attention to hospitality, by means of the pineapple symbol, as well as festivity, culture, and seasonality.

Product: Rolfs
Designer: Wayne Krimston
Design Firm: Murrie White Drummond & Lienhart Associates
Client: Amity Leather Products Company

An elegant, understated package was created for fashionable men's leather goods.

Product: Gift Box
Designer: Frans Lieshout
Design Firm: Total Design, Amsterdam
Client: Randstad Uitzendbureau bv

The use of contemporary perspective design and graphics creates an appealing gift box.

Product:	Wrapping Paper
Designer:	Alex Graham
Design Firm:	Total Design, Amsterdam
Client:	Randstad Uitzendbureau bv

Sharp colors and a contemporary, simple design are used for this gift wrapping.

Product:	Iveco Gift Wrap
Designer:	Paolo De Robertis
Design Firm:	PAN, Torino, Italy
Client:	Iveco, Torino, Italy

Iveco manufactures truck models for children. This wrapping paper follows suite, with small illustrations of trucks and related items. The design is appropriate for small boys' gifts.

Product: Weather's Men's Clothier
Designer: Semaphore Inc.
Design Firm: Semaphore Inc., Columbia, SC
Client: Weather's, Columbia, SC

A classy, aggressive, and masculine representation of men's clothing styles from Weather's retains a conservative format.

Product: Primo Angeli Gift Wrap
Designer: Staff, Primo Angeli Inc.
Client: Primo Angeli Inc., San Francisco, CA

This packaging idea allows for modular display as well as for functional year-round storage of wrapping papers and related items.

Chapter 6

Business Products

Computer accessories
Telephone accessories
Business electronic items
Video equipment

With products ranging from computer accessories to video equipment, this is truly a growth category for packaging designers. It is also a category where "good" design can be expressed because of the relative sophistication of the consumer and the fact that the package is designed to act as a source of information rather than a promotional billboard.

As is the case with housewares packaging, designers often use a systematic approach when arranging the information and/or diagrams that must appear on the packages of many business products. Many designers turn to basic grid design principles or Eurosytle design to create clear, organized designs that invite the consumer to examine the package.

Designers are meeting this new challenge and, at the same time, are creating professional-looking and graphically appealing packages that give the product a sophisticated, contemporary look. The packaging designs from Olivetti, consistent with the company's great design tradition, are outstanding, especially the designs for Olitext Plus (p. 186) and other accessories (pp. 192, 193), which are stylish and fully integrated. Even the shipping boxes are well-conceived: they provide clerks with a format for stocking the products that enables all important information to be easily read.

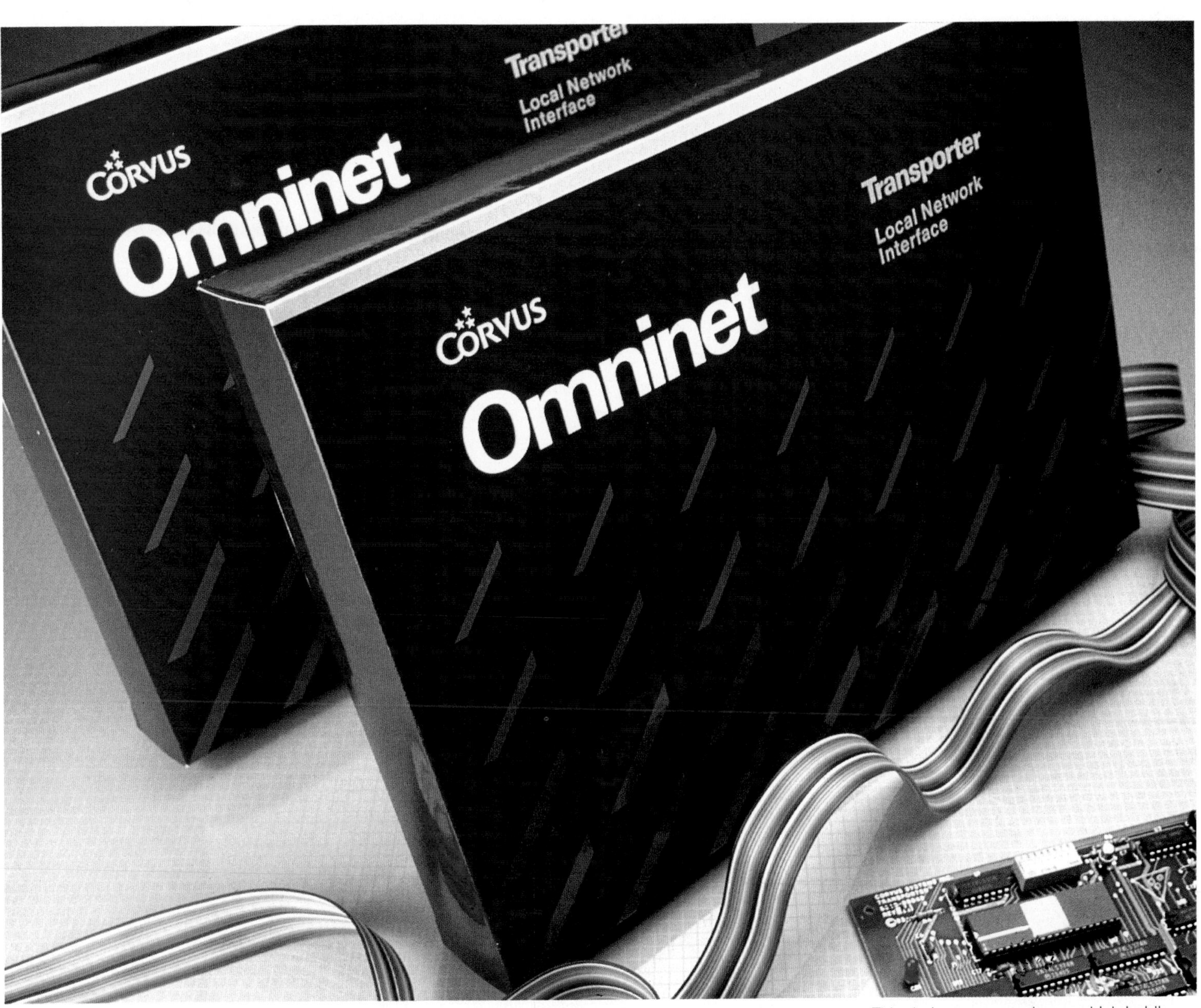

This dark maroon package, which boldly displays the product's brand name in white, conveys a sophisticated, professional image. Otherwise, the rich package is atypical of most contemporary business packaging, which tends to include a good amount of information.

Carolina Eng. Lab used this dynamic package design to launch Stylewriter, a product that enables any dot matrix printer to be converted into a letter-quality printer. Every element of the design is sleek and contemporary. While the package does not include a great deal of product information, as do those of some business products, it does present a concise product description.

Product: Olivetti Hard Disk Cards
Designer: Roberto Pieraccini
Design Firm: Olivetti Direzione di Corporate Image, Milano, Italy
Client: Olivetti Peripheral Equipment S.p.A., Torino, Italy

On this box of Olivetti Hard Disk Cards, the product is easily identifiable by a detailed package illustration.

Product: Telephone Answering Machine
Designer: Kathleen Campbell
Design Firm: Tom Campbell & Associates Inc., Los Angeles, CA
Client: Fortel/Record a Call, Compton, CA

Record a Call designed its packages for clarity and organization so that its products would be marketable in self-service mass merchandise outlets.

Product: Stylewriter
Designers: John Baker, M.C. Akers, and Tim Gilland
Design Firm: Design/Joe Sonderman Inc., Charlotte, NC
Client: Carolina Eng. Lab, Charlotte, NC

New to the computer trade, Stylewriter needed to gain immediate exposure. The product allows any dot matrix printer to be converted to letter quality. The sleek package and logotype were designed as a static animation of the conversion process.

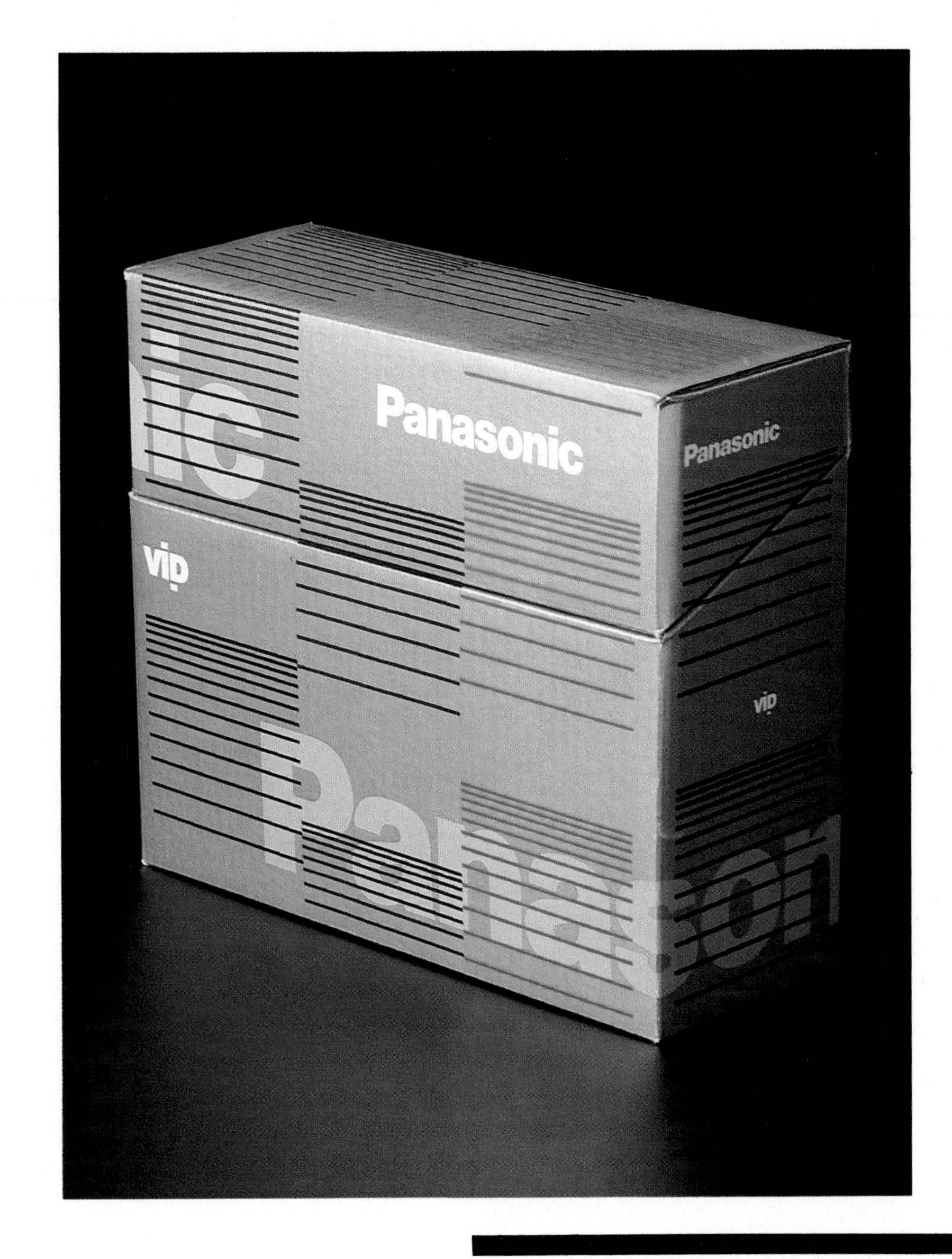

Product: Panasonic Video Information Program (VIP)
Designer: Gary Ludwig
Design Firm: The Spencer Francey Group, Ontario, Canada
Client: Matsushita Electric, Canada

Developed as part of a channel communications program, the cardboard flip-top box was used as a storage container/reference file for product literature and updates belonging to Panasonic video dealers.

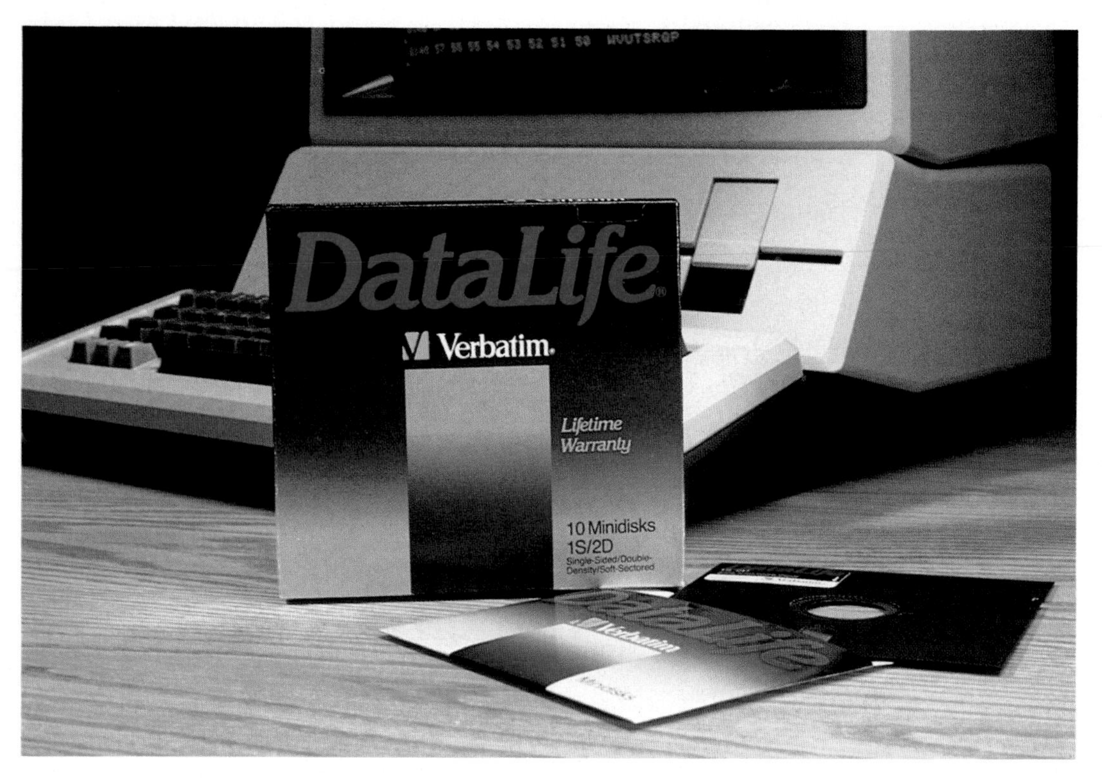

Product: Datalife Computer Diskettes
Designer: Edwin E. Ewry
Design Firm: S & O Consultants Inc., San Francisco, CA
Client: Verbatim Corporation, Sunnyvale, CA

This design has become a familiar sight to most computer users. The bold red brand name set against a navy background has been a contributing factor to this brand's popularity because it has created high visibility for the product.

Product: Software
Designers: Roberto Pieraccini and Antonia Fortarezza
Design Firm: Olivetti Direzione di Corporate Image, Milano, Italy
Client: Ing. C. Olivetti & C. S.p.A., Torino, Italy

A similar packaging design is carried across the Olivetti software line using computer graphics.

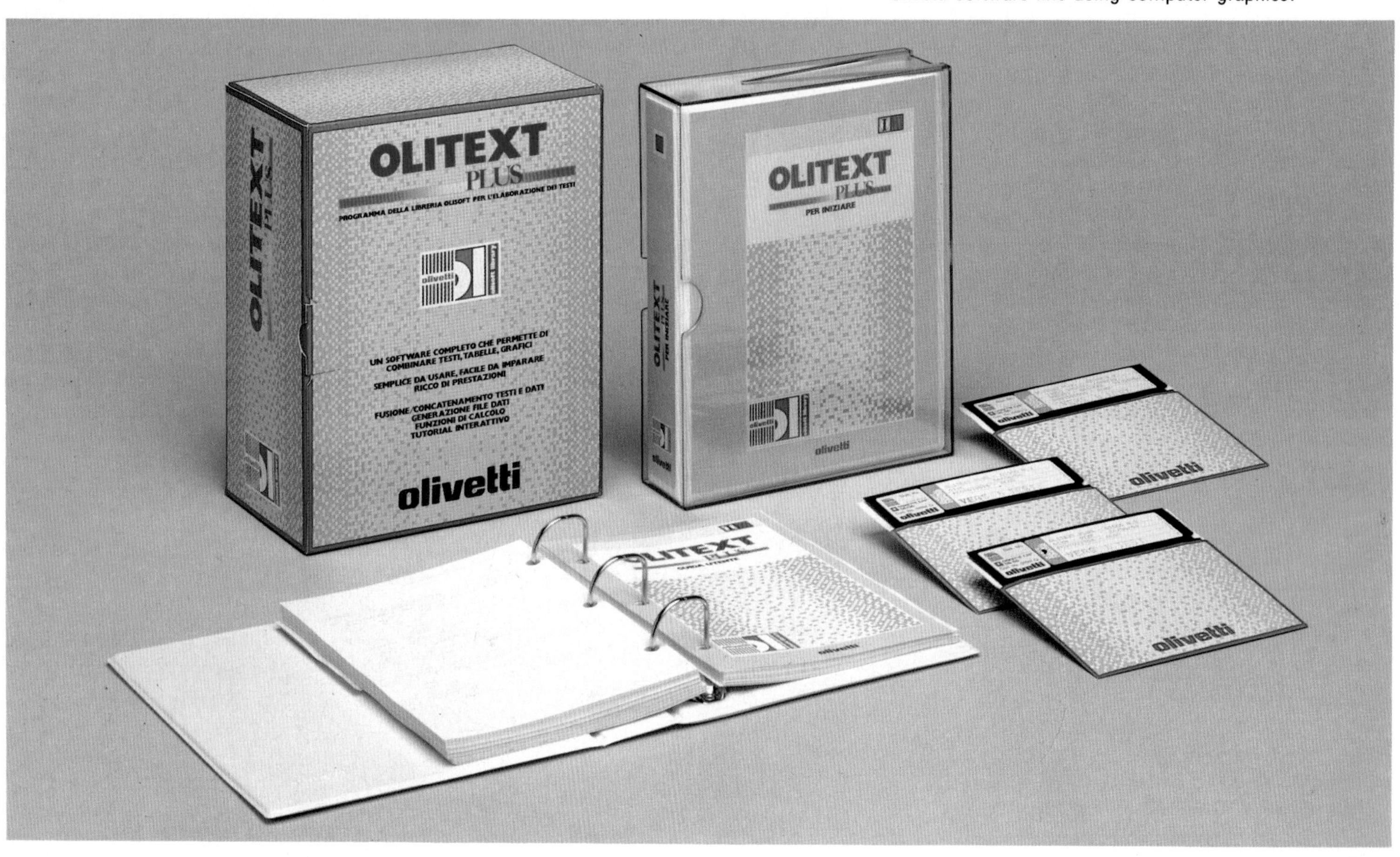

Product: Compact Disc Care-Pak
Designer: Linda Chow, N. Bogolubov, M. Isoldi, and Adam D'Addario
Design Firm: The Creative Source Inc., New York, NY
Client: Recoton Corporation

These products for compact disc players or VCR players are packed in blister packs with a unique attached carton. Die-cut windows are used on the Care Paks so that the products can "pop" through to show consumers what they are purchasing.

Product: Typewriter Software Package
Designer: Eric Anderson
Design Firm: IBM
Client: IBM, Lexington, KY

This packaging design was aimed at entry level software users. The packages present a straight-forward and uncomplicated reference to the product.

Product: Omninet Interface
Designer: Edwin E. Ewry
Design Firm: S & O Consultants Inc., San Francisco, CA
Client: Corvus Systems Inc., San Jose, CA

This packaging system reflects today's computer technology. A bold, clear brand name against a dark maroon panel allows for quick identification.

Product:	Cash Register and Personal Copier
Designer:	Group Four Design
Design Firm:	Group Four Design, Avon, CT
Client:	Royal Consumer Products

This is an extension of an established family look used previously for the company's typewriter and calculator packages.

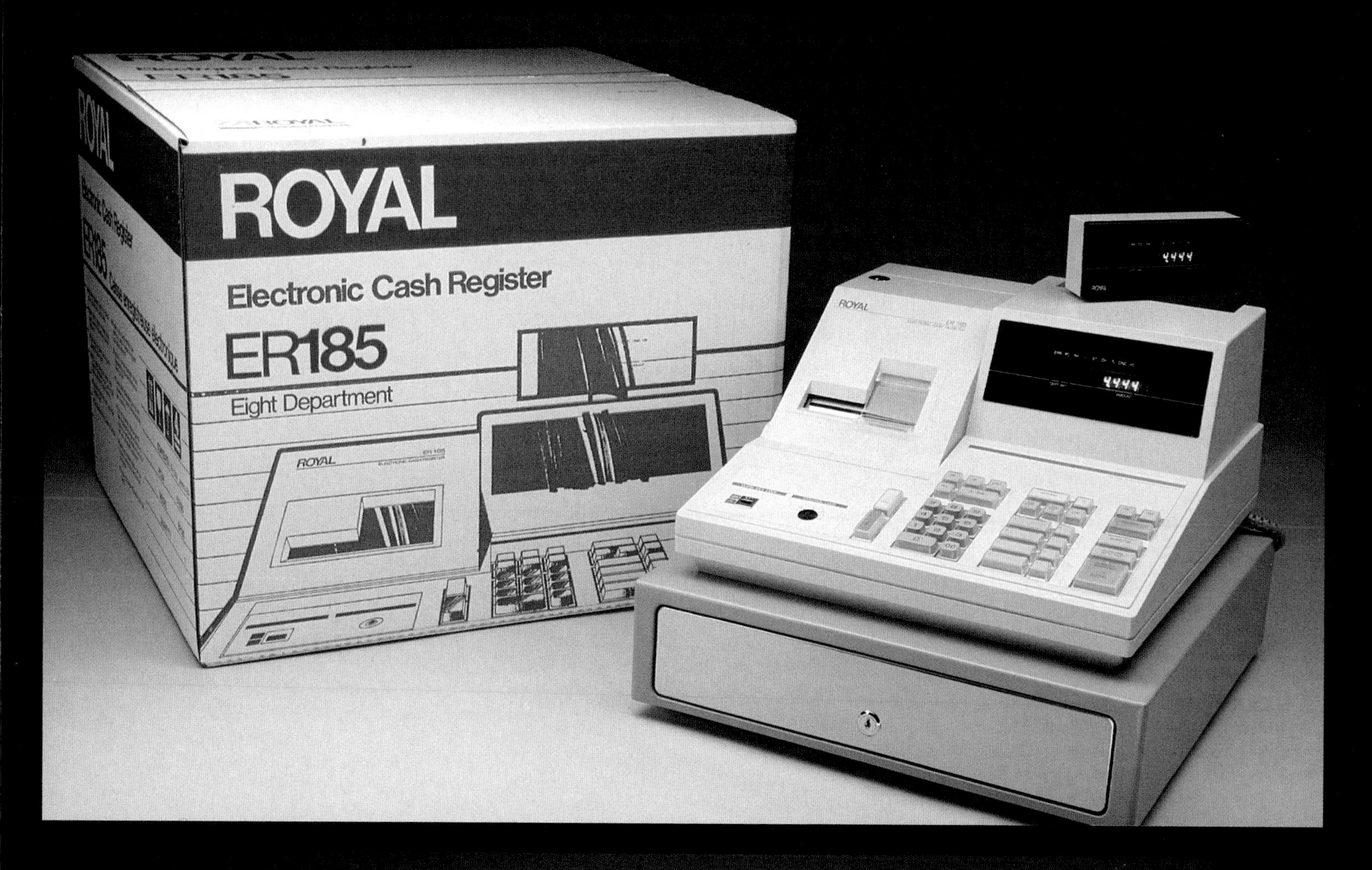

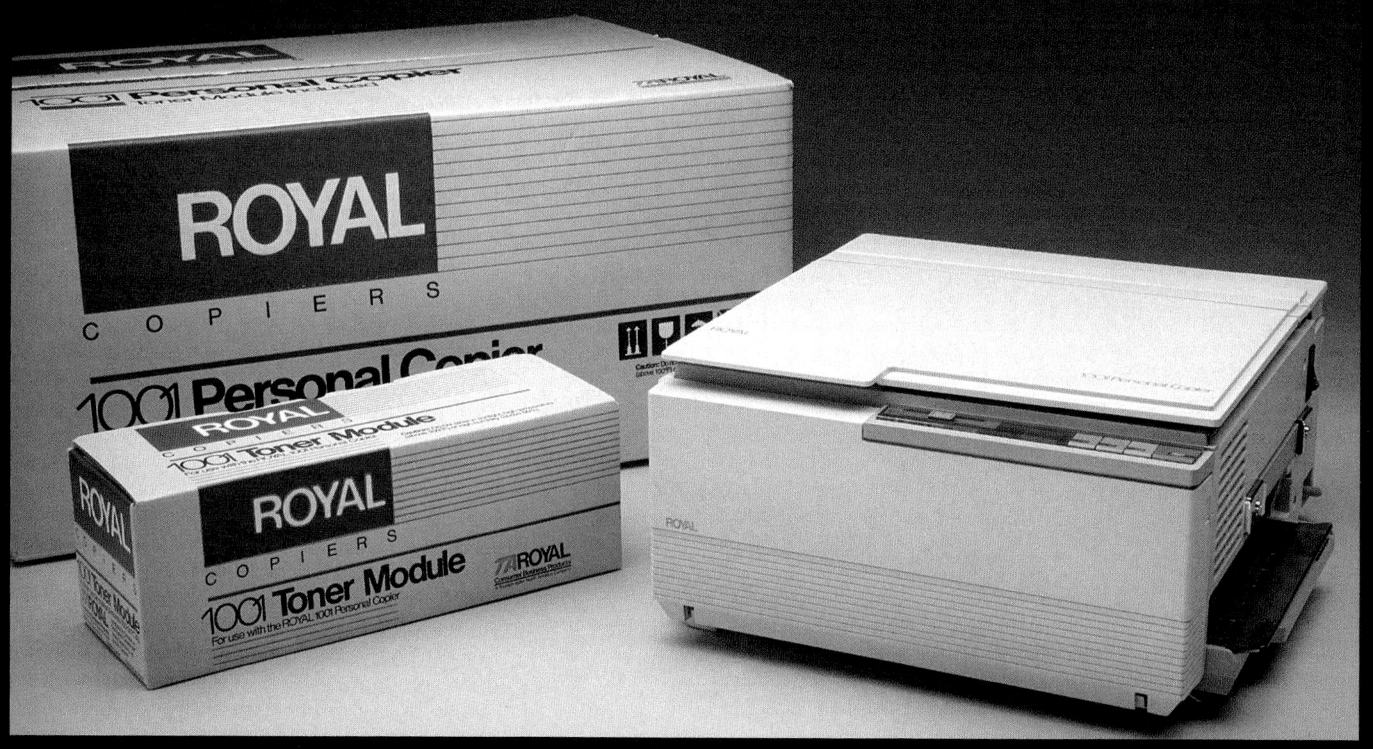

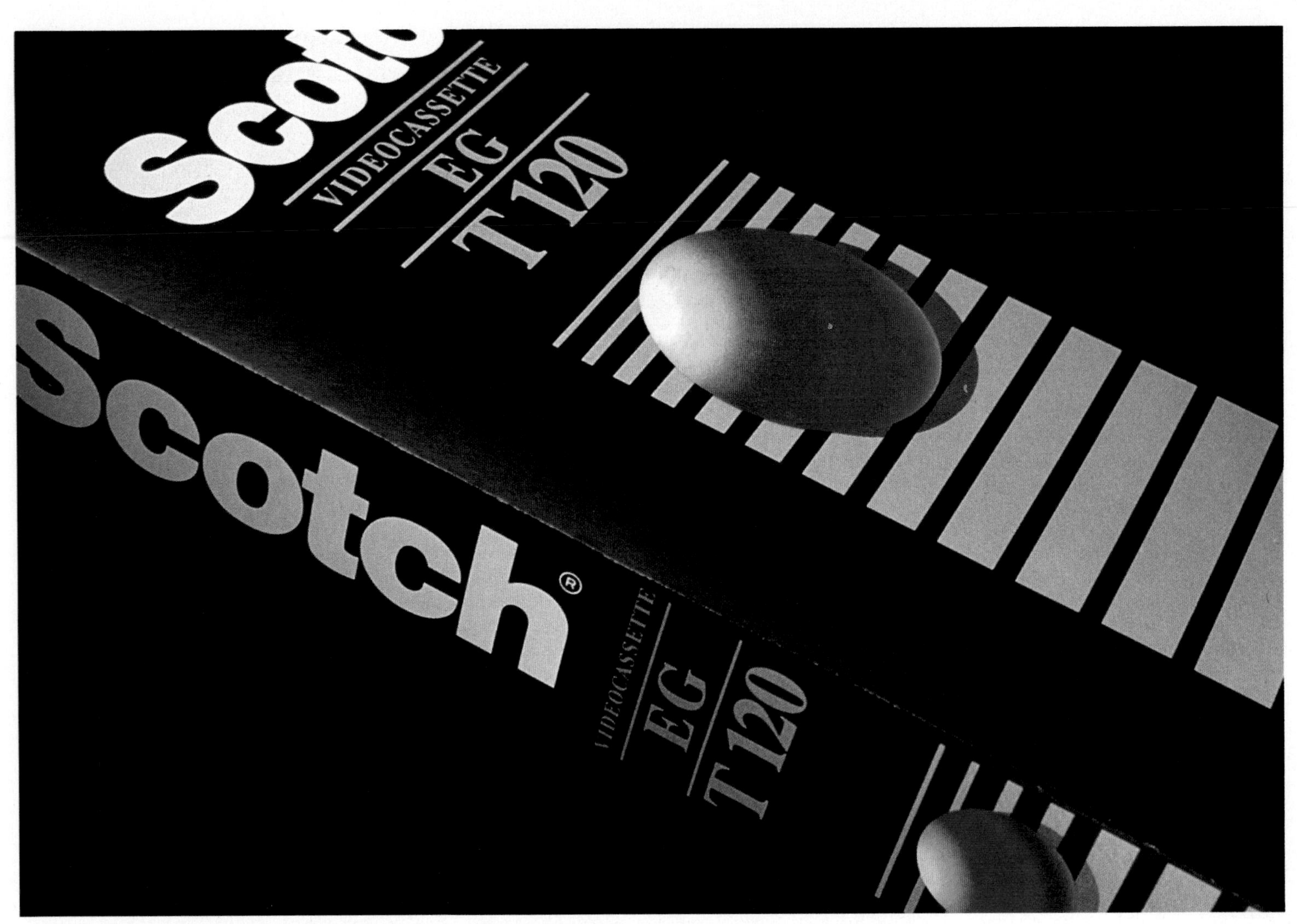

Product: Scotch Video Cassette
Designers: Albert Ng, Jack Vogler, and Kay Stout
Design Firm: Landor Associates, San Francisco, CA
Client: 3M, St. Paul, MN

The full-color spectrum sphere on the video cassette box conveys the product's major attribute—colorful, high-quality, state-of-the-art reproduction.

Product: Compucat Quizware
Designer: Group Four Design
Design Firm: Group Four Design, Avon, CT
Client: McGraw-Hill

Color-coded bands were used across this line of computer software to distinguish subjects. In addition, illustrations representing the subjects are printed on the front of the packages.

Product:	Wilson Learning II
Designers:	Hideki Vamamoto, Aimee Hucek, and Oscar Pena
Design Firm:	Seitz Yamamoto Moss Inc., Minneapolis, MN
Client:	Wilson Learning Company, Eden Prairie, MN

Interactive with today's computer technology, these video disk packages were designed with high tech and high style in mind. An anodized aluminum package was selected to give an exciting visual and tactile experience to the consumer.

Product: Memorex
Designer: Peggy Wong, Jack Vogler, and Kay Stout
Design Firm: Landor Associates, San Francisco, CA
Client: Memorex, Santa Clara, CA

This unified packaging system can be adapted for additions to the product line. It creates a unique personality image for a line of flexible discs and office supply products.

Product: Olivetti Accessories
Designer: Bob Noorda
Design Firm: Unimark, Milano, Italy
Client: Ing. C. Olivetti & C. S.p.A., Torino, Italy

These supplies for Olivetti typewriters are easily identifiable and are in line with Olivetti's corporate image.

Product: Olivetti Shipping Box
Designers: Hans von Klier, Clino Castelli, and Perry King
Design Firm: Olivetti Corporate Image Department, Milano, Italy
Client: Ing. C. Olivetti & C. S.p.A., Torino, Italy

The corporate logo on these packages must be clearly printed for easy identification at the warehouse.

Chapter 7

HARDWARE AND SUPPLIES

- INSECTICIDES
- PET CARE PRODUCTS
- MOTOR OILS
- ELECTRONICS
- CAMERA SUPPLIES/ACCESSORIES
- ART SUPPLIES

With everything from insecticides to art supplies, this is very much a catch-all category, in many respects a cousin to business products because of the intelligence brought to the design process. This is also a category where one finds real advances in basic packaging with the introduction of new materials and sculptural designs. Such is true of the new Mobil 1 oil package (p. 215) with its integrated handle and the IP motor oil package (p. 215) from Italy.

As is the case with the packaging for many houseware and business products, the packages for some hardware products and related supplies must present a good deal of information for the consumer. Often using basic grid design techniques or Eurostyle design, designers are meeting this requirement and creating exceptionally dynamic designs in the process. Packaging from Vivitar (pp. 206, 210, 212, and 213) is particularly well-handled: each of its packages, whether for an expensive item such as a 400mm camera lens or for one of its low-end consumer products, is intelligently planned and quite striking.

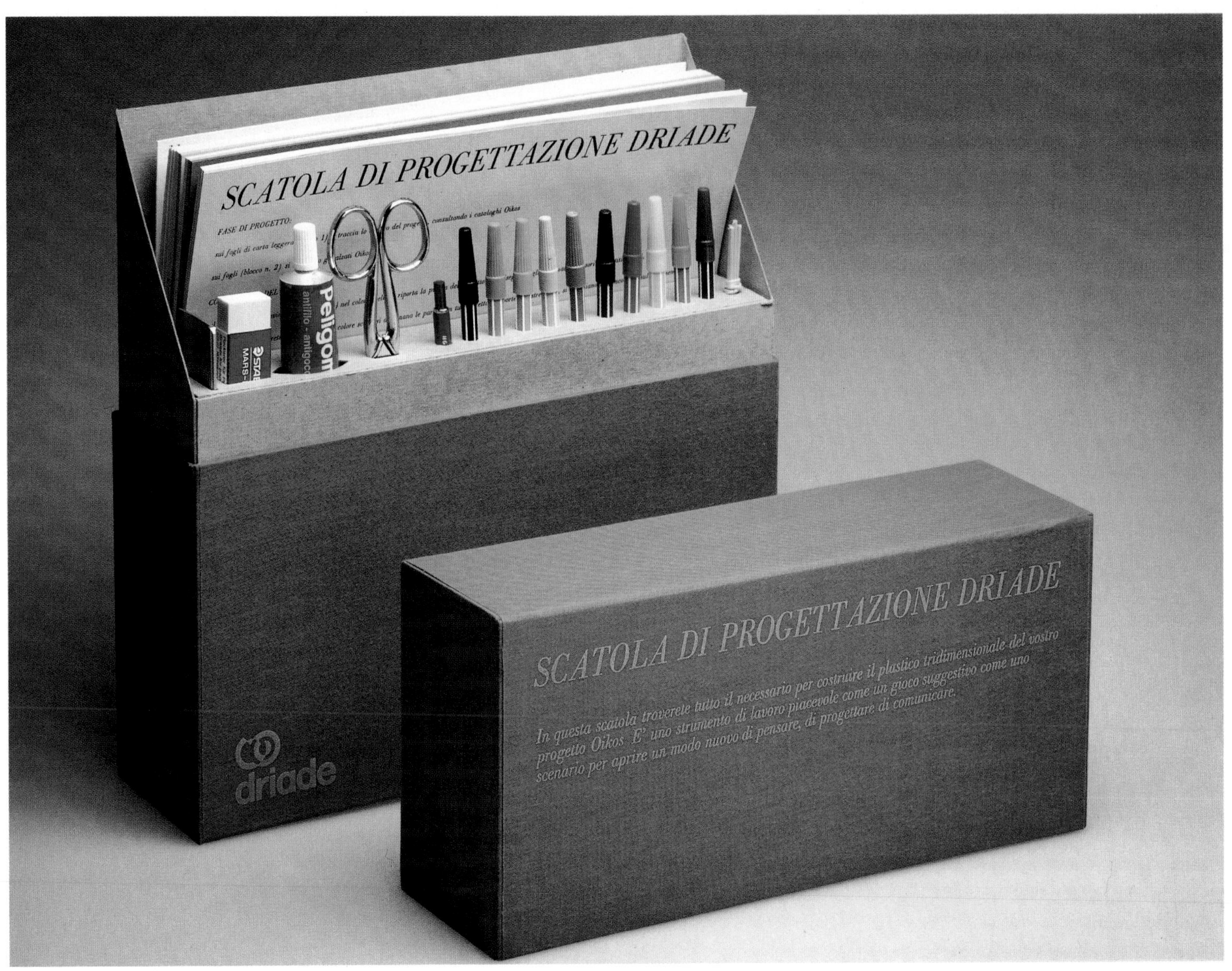

Created as a tool for the sales staff of Driade, a furniture company in Italy, this "building box" is exceptionally attractive, professional-looking, and perhaps most important, intelligently designed.

This container for Mobil 1 oil is one of the most striking examples of innovative sculptural packaging now in the marketplace. The integrated handle is intelligently planned and adds to the impact of the container's strong, sophisticated graphics.

Product: Gainseborough Introductory Oil Color Sets
Designers: Owen W. Coleman and Art Donovan
Design Firm: Coleman, LiPuma, Segal & Morrill Inc., New York, NY
Client: M. Grumbacher Inc., New York, NY

A bold, fun, contemporary feeling was created for this new line of introductory oil color sets by Grumbacher. The strong graphic image is considered a high selling point for mass-merchandising outlets nationwide.

Product: Visa Magic Markers
Designers: Paule Vezinhet and Pierre Le Gonidec
Design Firm: GE2A, Lille, France
Client: Conté, Paris, France

Designed to look like a jukebox, each of these magic marker sets contains 15 or 25 markers in a box.

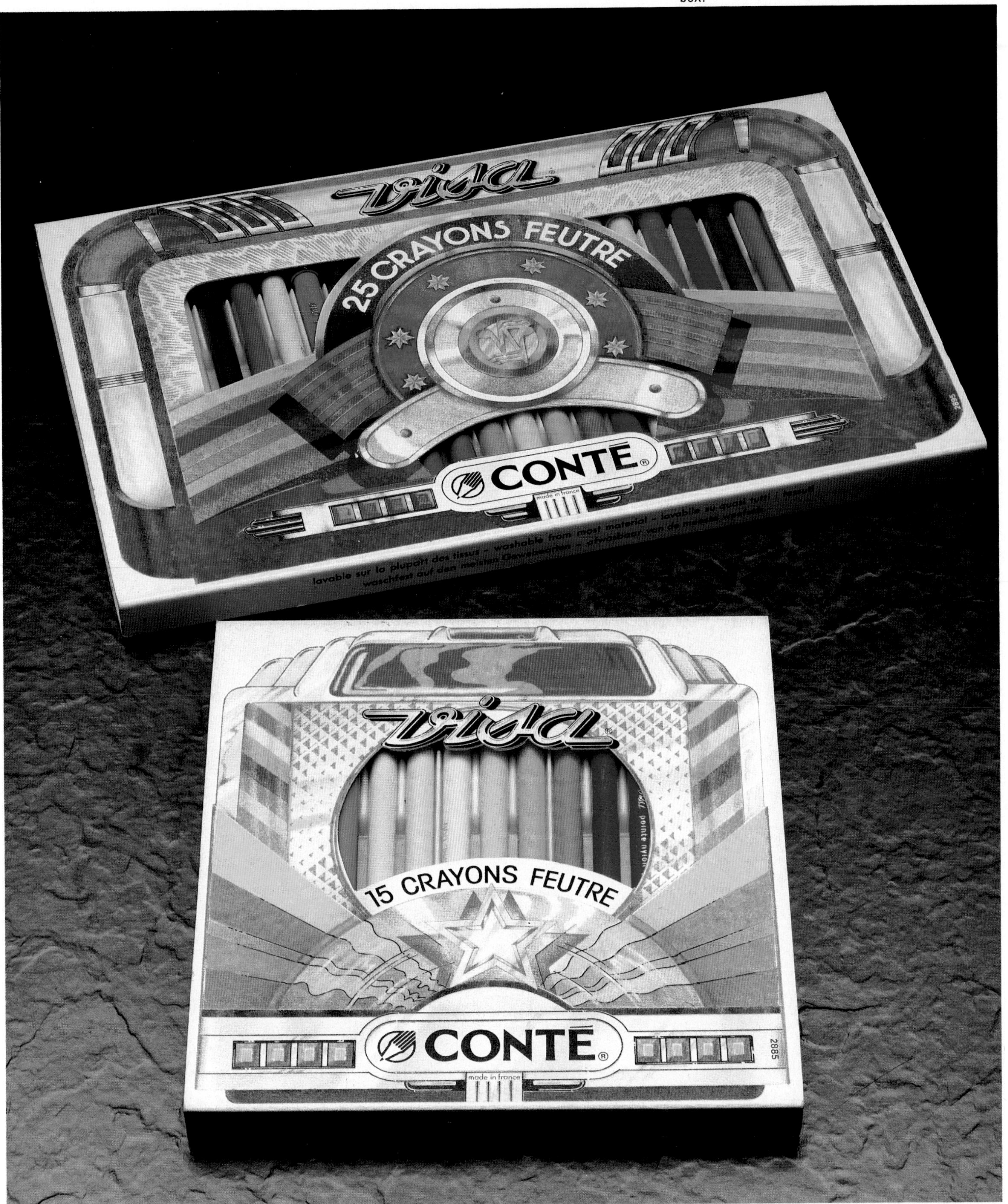

SINTONIE
LINEA DI SMALTI E IDROPITTURE IN SEI DIVERSE
TONALITA' DI BIANCO. ABBINABILI FRA LORO
i BIANCHI NATURALI
750 ml e
IDROPITTURA

MaxMeyer

Chicago, IL

Client: Lowe's Inc.

This new, stylized illustration and identifiable logo reflects Kitty Litter's acquired position in its field.

Product: IP Motor Oil
Designer: Luciano Lorenzi
Design Firm: G & R Associati, Milan, Italy
Client: Sintiax, Milan, Italy

This Sintiax product is packaged in a tall containe that is easy to manuever under a car hood. The IP symbol appears to be in motion, signifying a gooc quality motor oil.

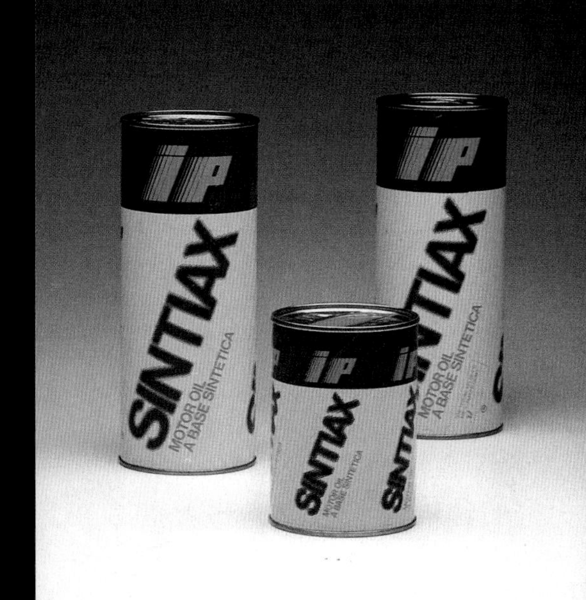

nie Paints
Diagonale Advertising Srl
Diagonale Advertising Srl, Milano, Italy
Meyer Duco S.p.A., Milano, Italy

ifferent tones of white, Sintonie r-base paints are labelled the Soft pastel coloring against a ground was chosen to represent the ge.

Product: Building Box
Designer: Adelaide Acerbi
Design Firm: Lambda, Milano, Italy
Client: Driade, Milano, Italy

This building box contains all the necessary pieces to build a model of any interior. The box is given to the sales staff of Driade, a furniture company.

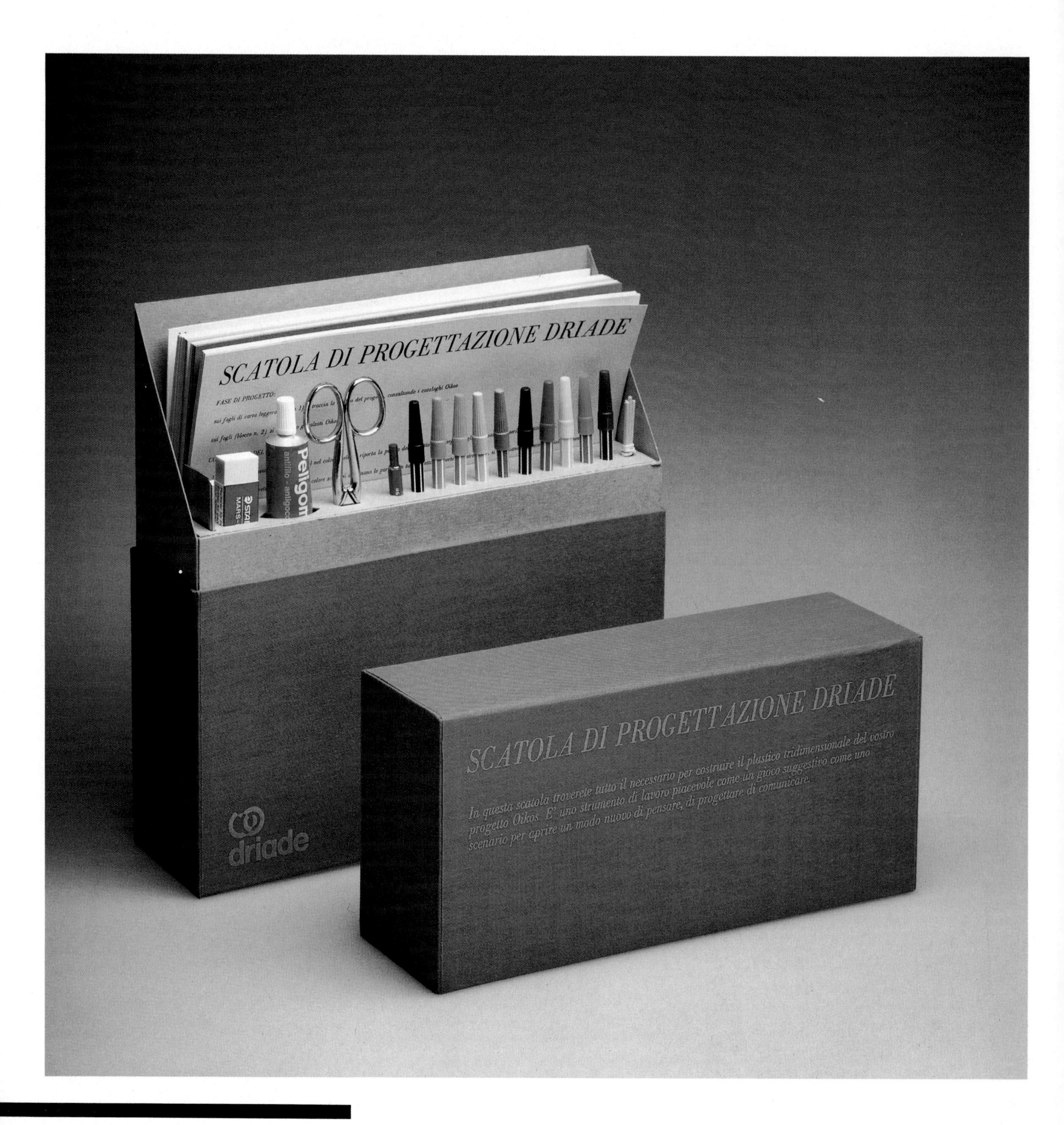

Product: Castrol Motor Oils
Designers: Sal LiPuma and William Lee
Design Firm: Coleman, LiPuma, Segal & Morrill Inc., New York, NY
Client: Burmah-Castrol Inc., Hackensack, NJ

A plastic container with a new shape and bold contemporary graphics were used for a variety of motor oil products.

Product: Amoco Oil
Designer: Joe Selame
Design Firm: Selame Design, Newton, MA
Client: Amoco Oil Company, Chicago, IL

Graphic and color representation create an easily recognized package for Amoco, continuing a strong family identity for the large product line.

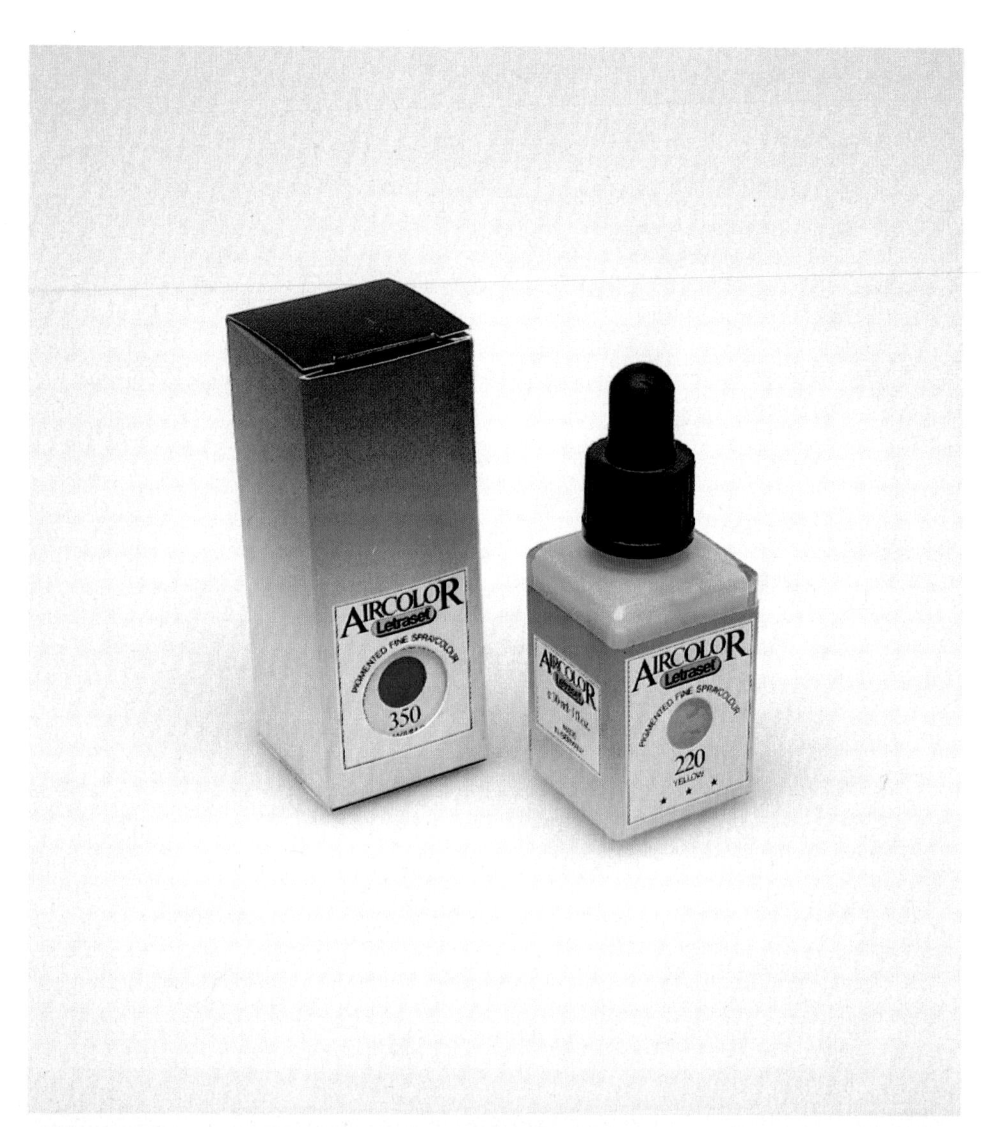

Product: Aircolor by Letraset
Designer: Emilio Fioravanti
Design Firm: G. & R. Associati, Milano, Italy
Client: Letraset, Italy

This attractive and convenient package for an air brush ink shows the actual color through a window.

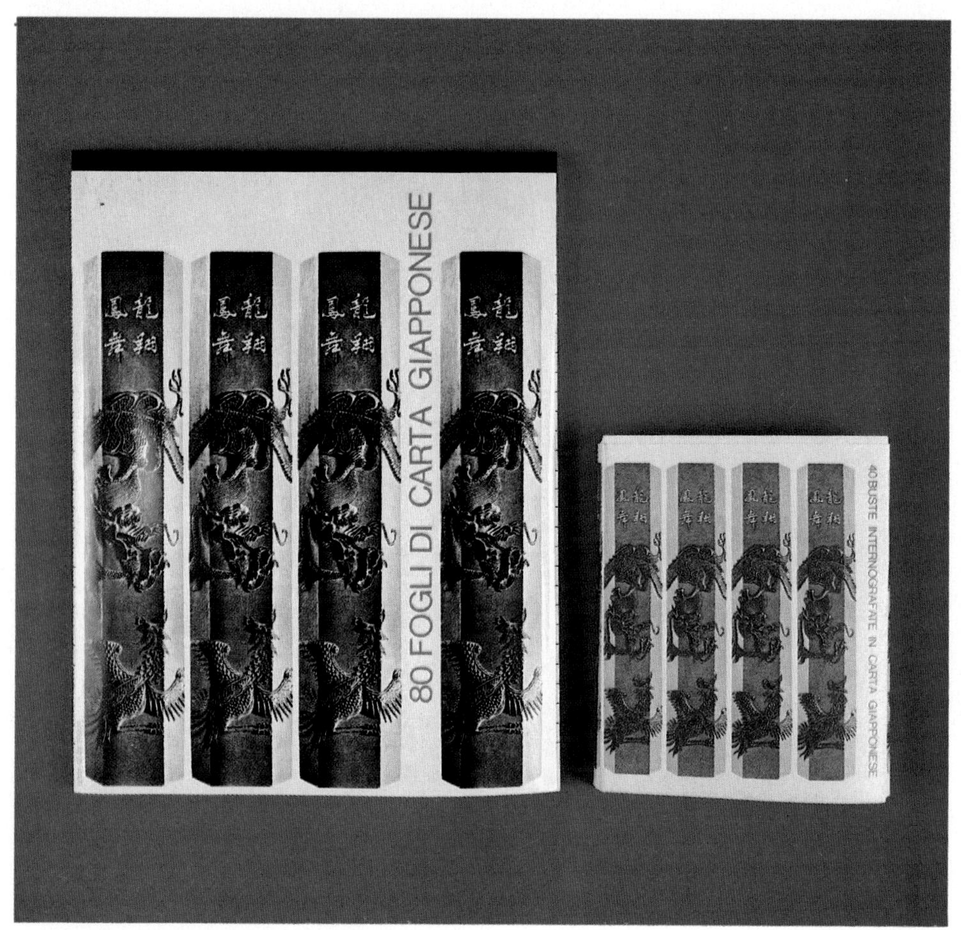

Product: Stationery
Designer: Ennio Lucini
Design Firm: Studio Elle, Milano, Italy
Client: La Rinascente and Upim Department Stores, Italy

This elaborate sketch of a dragon and a phoenix provides an interesting package illustration for stationery that is manufactured exclusively for the La Rinascente and Upim department stores. The illustrations reflect those imprinted on bricks of solid ink sold in China.

Product: Drawing Kits
Designer: Ennio Lucini
Design Firm: Studio Elle, Milano, Italy
Client: La Rinascente and Upim department stores

These kits were designed for two department stores, La Rinascente and Upim. Each item is assigned its own space in the package for a uniform, organized format.

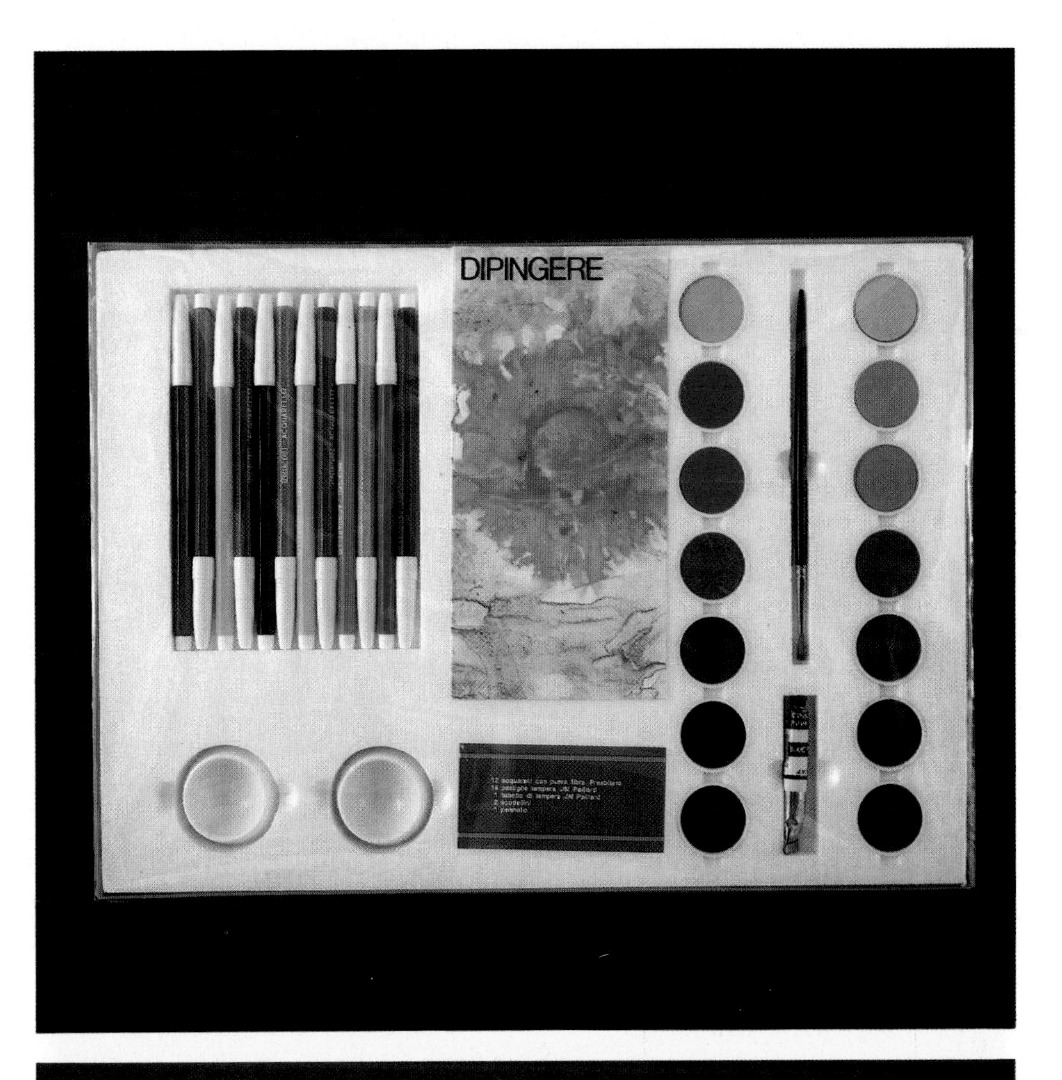

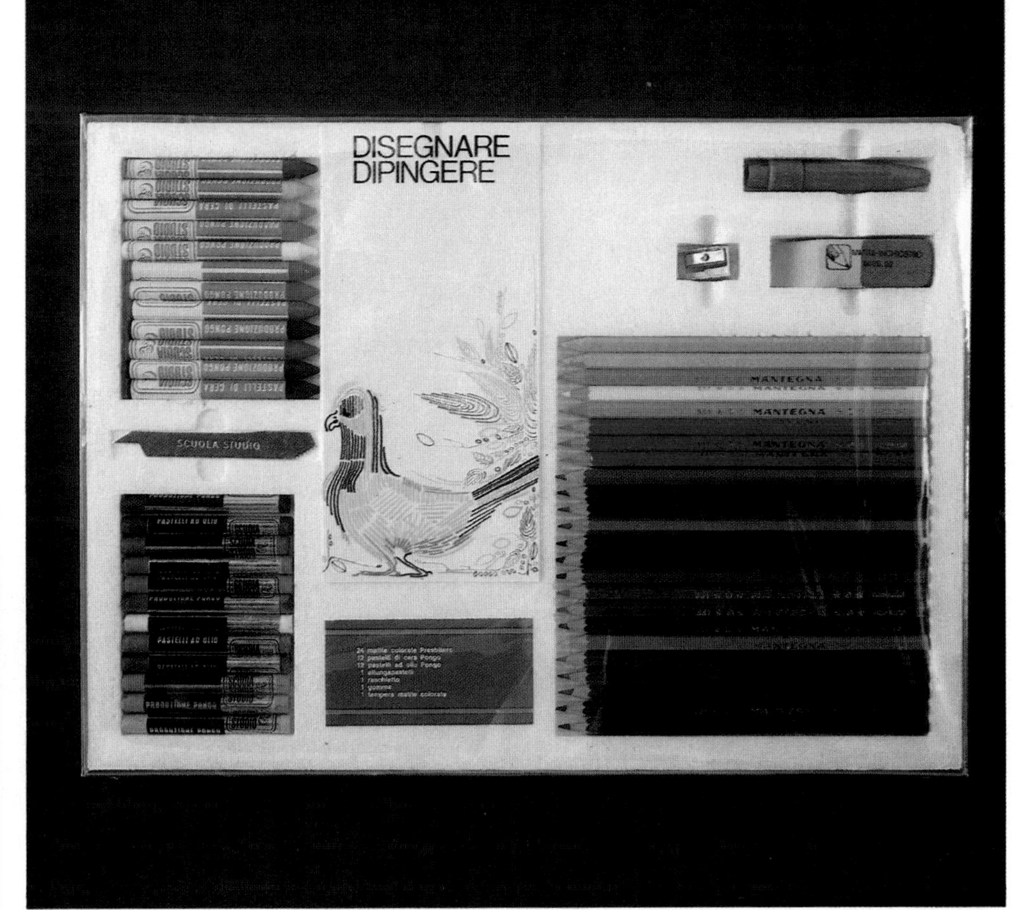

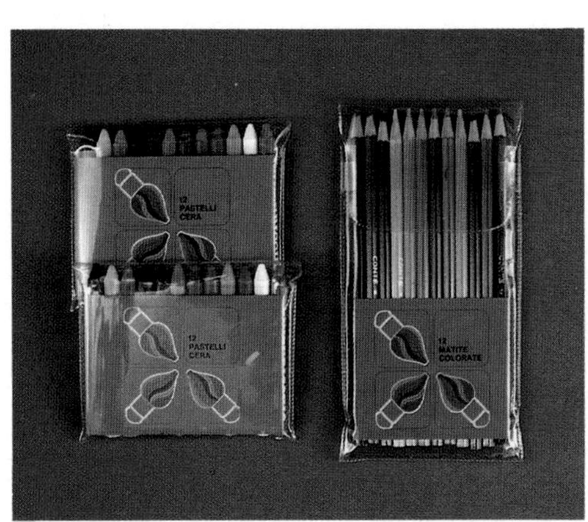

Product: Raid Insecticide Products
Designer: Kornick Lindsay
Design Firm: Kornick Lindsay, Chicago, IL
Client: S.C. Johnson

The packages for three major brands were redesigned to increase shelf impact for the line while differentiating each product by specific use.

Product: Sint 2000 Motor Oil
Designer: Gió Rossi
Design Firm: Image Plan International, Milano, Italy
Client: Agip, Rome, Italy

Graphics and colors combine to form a new, contemporary image for Sint 2000 Motor Oil.

Product: Nu-Hue Spray Paint
Designer: Edward Rebek
Design Firm: John Racila Associates, Oak Brook, IL
Client: Dupli Color Products Company, Elk Grove Village, IL

The packaging design for this new line of spray paint creates strong impact and a quality impression. The clean verticle format dramatizes the can's height, and the use of color gradation suggests the range of colors offered in the product line.

Product: Vivitar System 35
Designers: Alan Stone and Cyndi Brooks
Design Firm: Alan Stone Design, Sherman Oaks, CA
Client: Vivitar Corporation, Santa Monica, CA

This packaging system is designed to hold equipment for the beginner photographer. The back panel shows some sample photos.

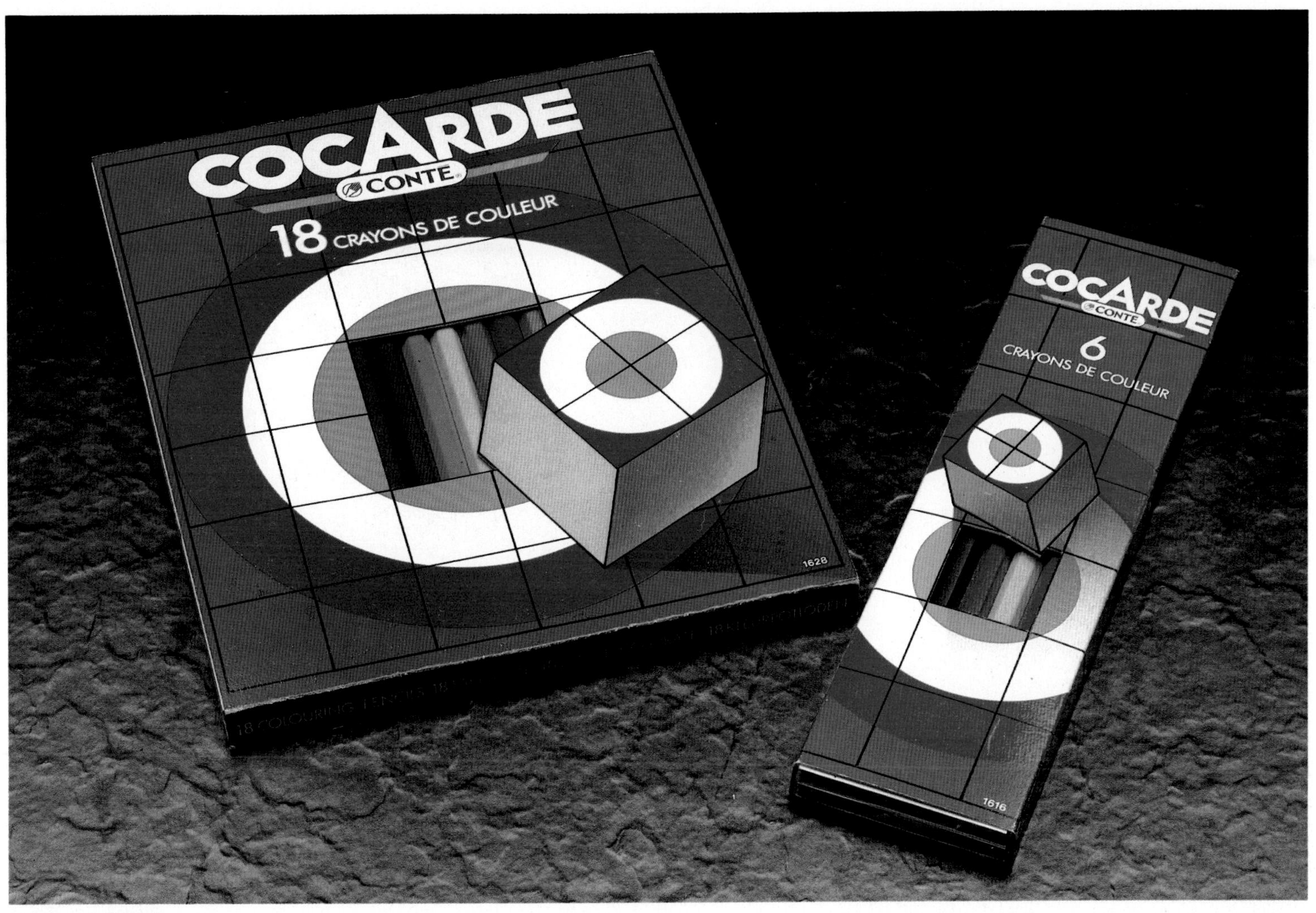

Product: Cocarde
Designers: Paule Vezinhet and Pierre Le Gonidec
Design Firm: GE2A, Lille, France
Client: Conté, Paris, France

This coloring pencil set illustrates a three-dimensional perspective on a two-dimensional surface.

Product: Log Lighter
Designer: Dixon & Parcels Associates
Design Firm: Dixon & Parcels Associates, New York, NY
Client: Tiger Tim Distribution Company, Inc., Wayne, PA

This product is highlighted on the face of the package by an actual photograph of Log Lighter starting a fire. The product name dominates in white against a dark background. The back of the package provides easy-to-use instructions.

Product: Konica Film
Designers: S & O Design Team, Edwin E. Ewry
Design Firm: S & O Consultants Inc., San Francisco, CA
Client: Konica Corporation, Englewood Cliffs, NJ

Panasonic employed new product graphics and a new packaging design for its battery cell and flashlight lines, which were recently introduced to the American market.

Product: Quaker State Oils
Designer: Dixon & Parcels Associates
Design Firm: Dixon & Parcels Associates, New York, NY
Client: Quaker State Oil and Refining Company, Oak City, PA

A complete redesign for Quaker State was executed, including a new, foil-based label and a new "Q" with a color-coded band for product identification.

Product: FOB Sports Ammunition
Designer: Hotshop
Design Firm: Hotshop, Paris, France
Client: S.N.P.E., Paris, France

The Michel Carrega shells are conveniently packaged in two boxes containing five shells each. The orange "FOB" name on the box with a bulls-eye effect, combined with the gold Michel Carrega signature, is prominent.

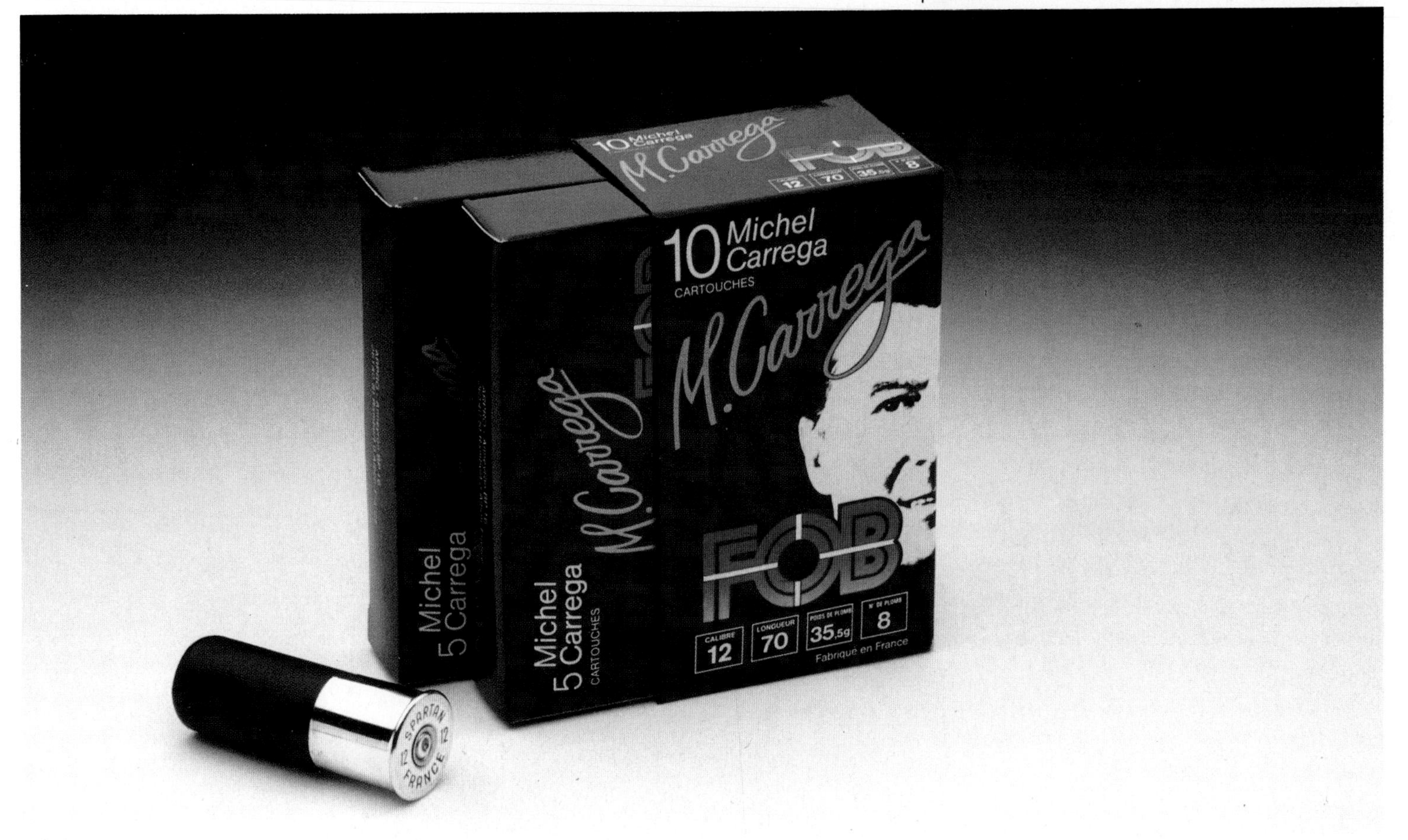

Product:	Compact 35mm Camera Kits
Designer:	Kathleen Campbell
Design Firm:	Tom Campbell & Associates Inc., Los Angeles, CA
Client:	Vivitar Corporation, Santa Monica, CA
Award:	"The Best in Packaging, *Print Casebook 7*

The front panel of each "travel kit" package functions like a billboard in the store, using bold simple graphics for impact, while the back panel depicts the content as well as information about the product.

Product: Sleek
Designer: Amy Leppert
Design Firm: Murrie White Drummond & Lienhart Associates, Chicago, IL
Client: S.C. Johnson & Son Inc.

The distinctive silver Sleek logo and "slipstream" graphic help position this product in the car wax market.

Product: Panasonic Flashlights
Designer: Group Four Design
Design Firm: Group Four Design, Avon, CT
Client: Panasonic

Panasonic employed new product graphics and a new packaging design for its battery cell and flashlight lines, which were recently introduced to the American market.

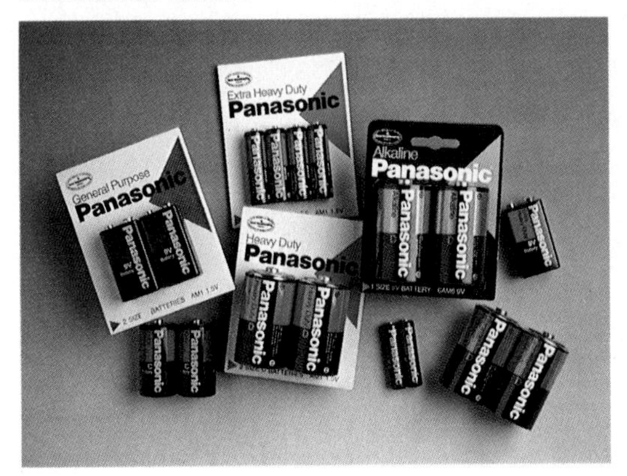

Product: Electronic Flash and Accessories
Designer: Kathleen Campbell
Design Firm: Tom Campbell & Associates Inc., Los Angeles, CA
Client: Vivitar Corporation, Santa Monica, CA
Award: "The Best in Packaging"

The packaging for these camera accessories was designed to convey a strong, consistent corporate image as well as to communicate important features in four languages (for worldwide distribution). Products are clearly differentiated by a stylized line drawing and bold product code.

Product: Compact Cameras
Designers: Tom Campbell and Kathleen Campbell
Design Firm: Tom Campbell & Associates, Los Angeles, CA
Client: Vivitar Corporation, Santa Monica, CA

The clamshell packages were designed for marketing compact cameras in self-serve mass merchandise and discount stores. A family theme was used to create a strong presence for Vivitar cameras on the display rack while color-coding was used to differentiate the cameras. The back is constructed as an easel to enable the packages to stand by themselves.

Product: Vivitar Series 1
Designer: Kathleen Campbell
Design Firm: Tom Campbell & Associates Inc., Los Angeles, CA
Client: Vivitar Corporation, Santa Monica, CA
Award: 1985 American Institute of Graphic Arts award

The high-tech cutaway illustrations on the Vivitar Series 1 packages are used to convey the precision craftsmanship; of the lens line. The bright pink and orange numeral one is a prominent identifier on the package.

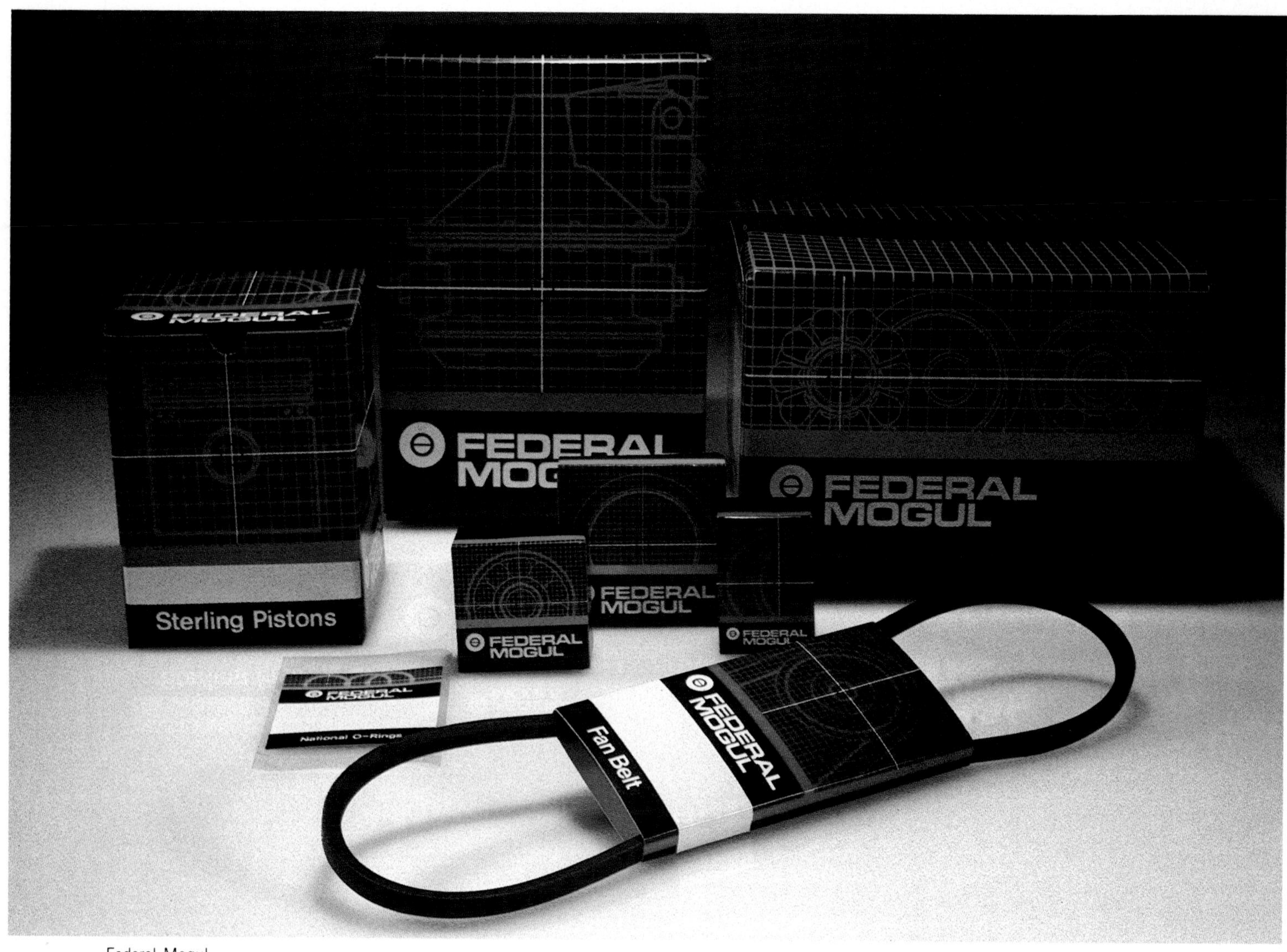

Product: Federal-Mogul
Designer: Adell Crump
Design Firm: Dickens Design Group, Chicago, IL
Client: Federal-Mogul Corporation, Southfield, MI

Packages for Federal-Mogul's bearings, seals, pistons, pumps, and carburetors are color-coded to ensure distributor access to computer inventory management techniques. The result is faster and more centralized delivery.

Product: Quicksilver
Designers: Ed Illig, Lisa Illig, Nancy Chifala, and Carolyn Zudell
Design Firm: Design Forum Inc., Dayton, OH
Client: Mercury Marine, Fond-du-Lac, WI

Packaging strategy included keeping a uniform, consistent look throughout the Quicksilver line.

Product: Mobil 1
Designer: Beautiful Designs
Design Firm: Beautiful Designs, Paris, France
Client: Mobil Oil

A unique package design upgrades the image of Mobil's high-grade oil and distinguishes it from other items in the product line.

Product: IP Motor Oils, Anti-freeze, Pontiax
Designer: Gió Rossi
Design Firm: Image Plan International, Milano, Italy
Client: Industria Italiana Petroli-IP, Milano, Italy

A new, modern look was created for IP products and a consistent, uniform design was maintained throughout the line.

Chapter 8

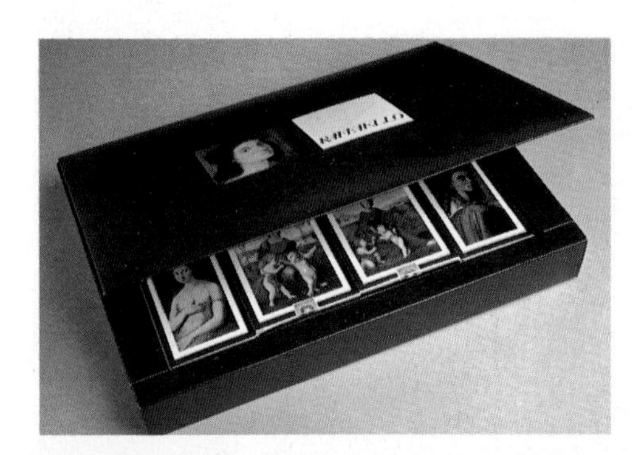

Miscellaneous Products

- Wearables
- Cigarettes
- Toys
- Games
- Sporting goods
- Jewelry

This final chapter is quite a potpourri. It includes packaging from a variety of product categories: wearables, cigarettes, books, toys, games, sporting goods, and jewelry.

Among the most interesting packages are those for Gitano hosiery (p. 236). Colorful, dynamic, and youthful, these packages of varying widths are not only striking but highly functional as well. A bright color code identifies the hosiery type within and a die-cut window invites the consumer to examine the actual product.

Another exceptionally well-designed package comes from Giorgio Armani (p. 237). Created by Minale, Tattersfield and Partners, this uniquely shaped, two-part container slides open. When the award-winning package is closed, a triangular window reveals the wearable inside.

One final package design deserves special mention: that created for the Daiwa microcomputerized fishing reels (p. 232) by Tom Campbell & Associates Inc. Highly technical drawings illustrate and identify product features while the precision of the renderings infers maximum performance.

Black or deeply colored graphic symbols, set against a quiet-colored background, impart a sense of intellectual sophistication to these books on the creative arts.

Designed to create a striking display of vibrantly colored triangles when stocked on the store shelf, the uniquely shaped boxes for Kenso Album tights present only four words. This maximizes the visual impact of the striking package colors and the dynamic package shape.

Product: Iveco Truck Models
Designers: Paolo De Robertis and Giovanna Ceste
Design Firm: PAN, Torino, Italy
Client: Iveco, Torino, Italy

A contemporary package shows the truck models "in motion," as "Champions of the Road." Floating in the sky above the trucks are various items that can be transported by truck.

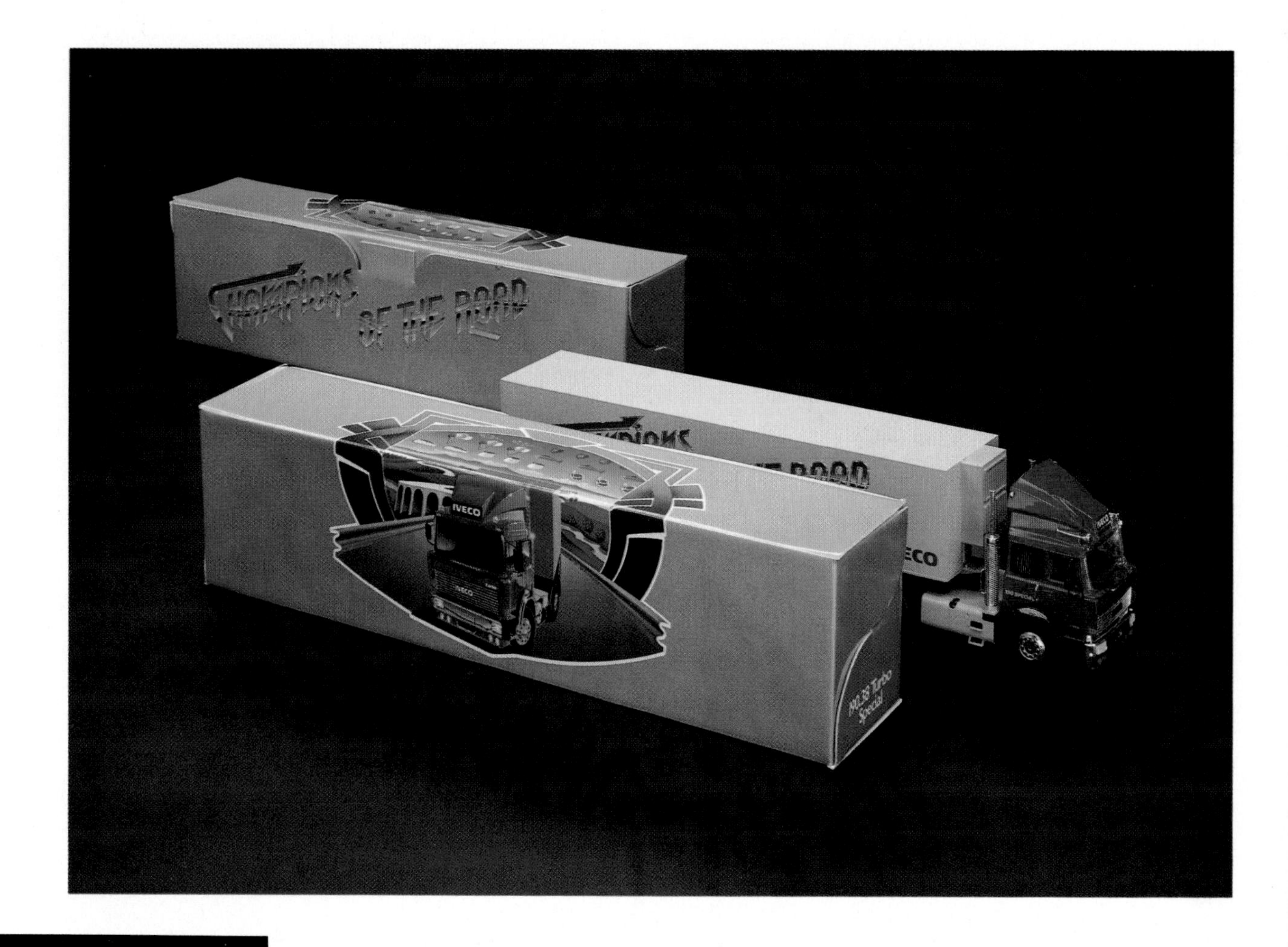

Product: Harrods/French Packaging
Designer: Minale, Tattersfield & Partners Limited
Design Firm: Minale, Tattersfield & Partners Limited, London, England
Client: Harrods Department Stores, London, England

For a recent special in-store promotion for "French Week," Harrods of London commissioned Minale, Tattersfield & Partners to design a wide range of packaging for French foods on sale in the store, among them tinned goods, bottles of vinegar, and jars of jam, herbs, and spices.

Product: Magnetix Design Set
Designers: Joe Selame, Richard Edlund, and David Adams
Design Firm: Selame Design, Newton, MA
Client: American Publishing, Watertown, MA

This package for the Magnetix Design Set is geared towards achieving high shelf recognition while complementing other packaging in the same line.

Product: Jem/Jerrica
Designer: Group Four Design
Design Firm: Group Four Design, Avon, CT
Client: Hasbro Inc.

Explosive graphic design for this product and its accessories helps attain shelf recognition and youth appeal.

Product: Album Tights
Designers: Peter Petronio and Xavier de Bascher
Design Firm: Concept Groupe, Paris, France
Client: Kenzo

Kenzo set out to design an original box shape for Album tights. The packages are printed in a variety of colors to create an explosion of triangular shapes when stocked on the store shelf.

Product: My Buddy
Designer: Group Four Design
Design Firm: Group Four Design, Avon, CT
Client: Hasbro Inc.

The packaging for this product complements other packaging in the line.

Product: Matchboxes
Designer: Emilio Stucchi
Design Firm: Staff, Saffa, Milano, Italy
Client: Saffa, Milano, Italy

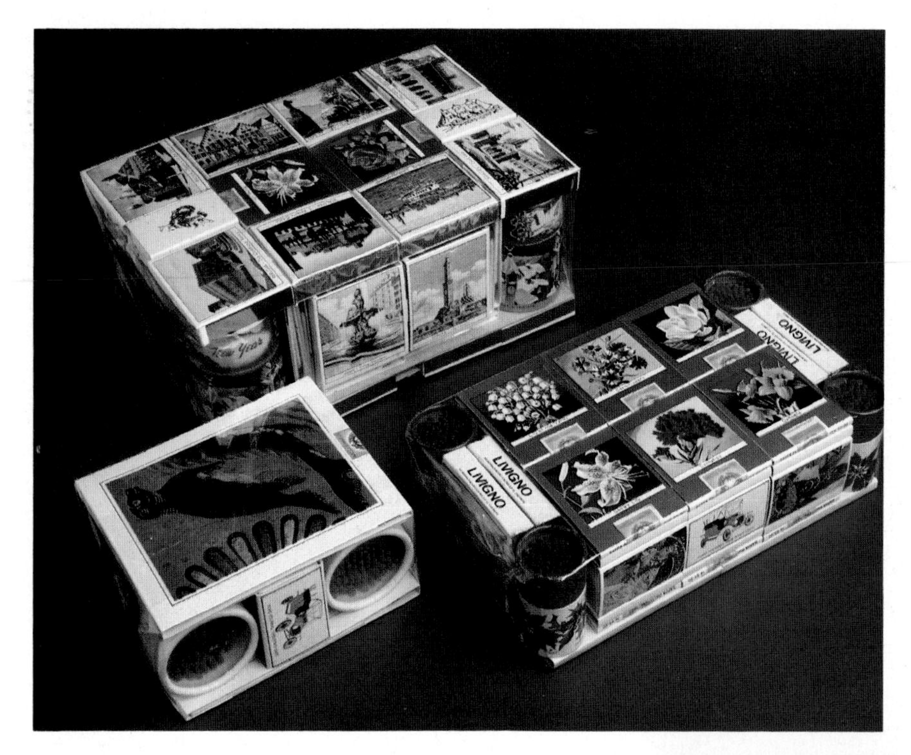

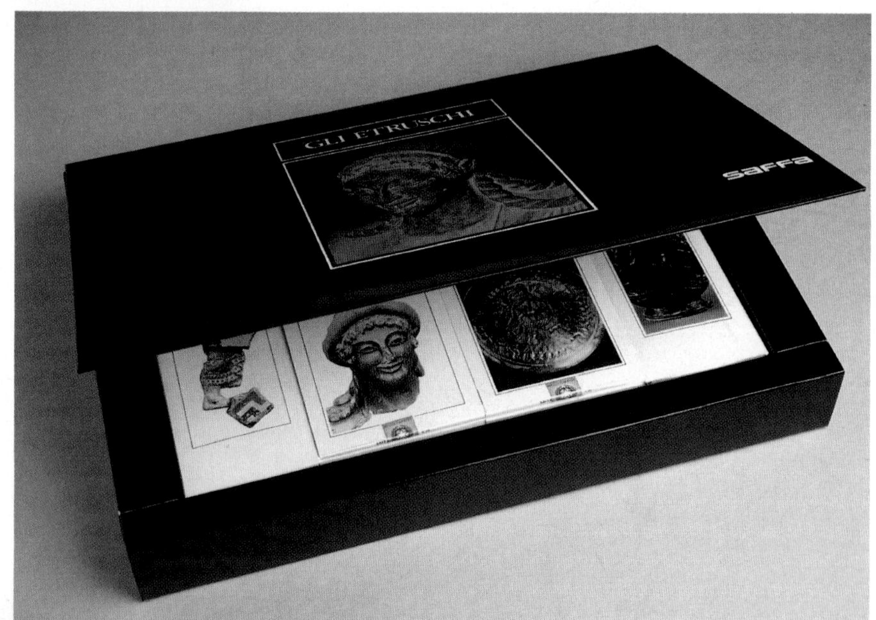

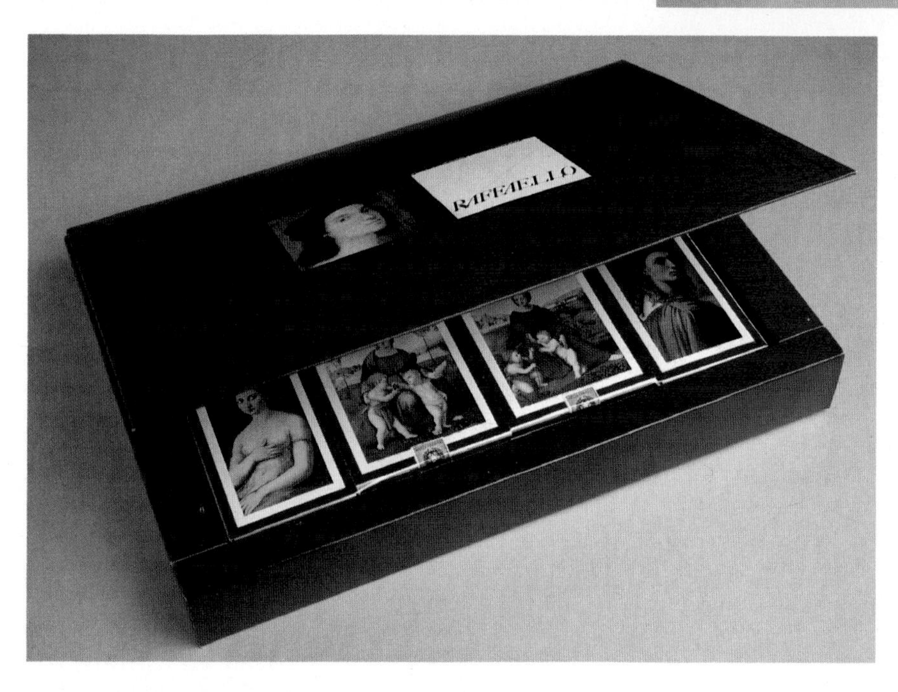

Matchboxes from Italy's major producer of matches and related products, Saffa, are shown here. The automobile series had a limited production, and the collection dedicated to Raffaello, as well as the Etruschi series, are not for sale.

Product: LTB
Designer: Thomas Q. White
Design Firm: Murrie White Drummond & Lienhart Associates, Chicago, IL
Client: Lorillard

This was a new entry in the 120mm cigarette category. The package creates a stylish, masculine image and suggests a rich, natural tobacco flavor.

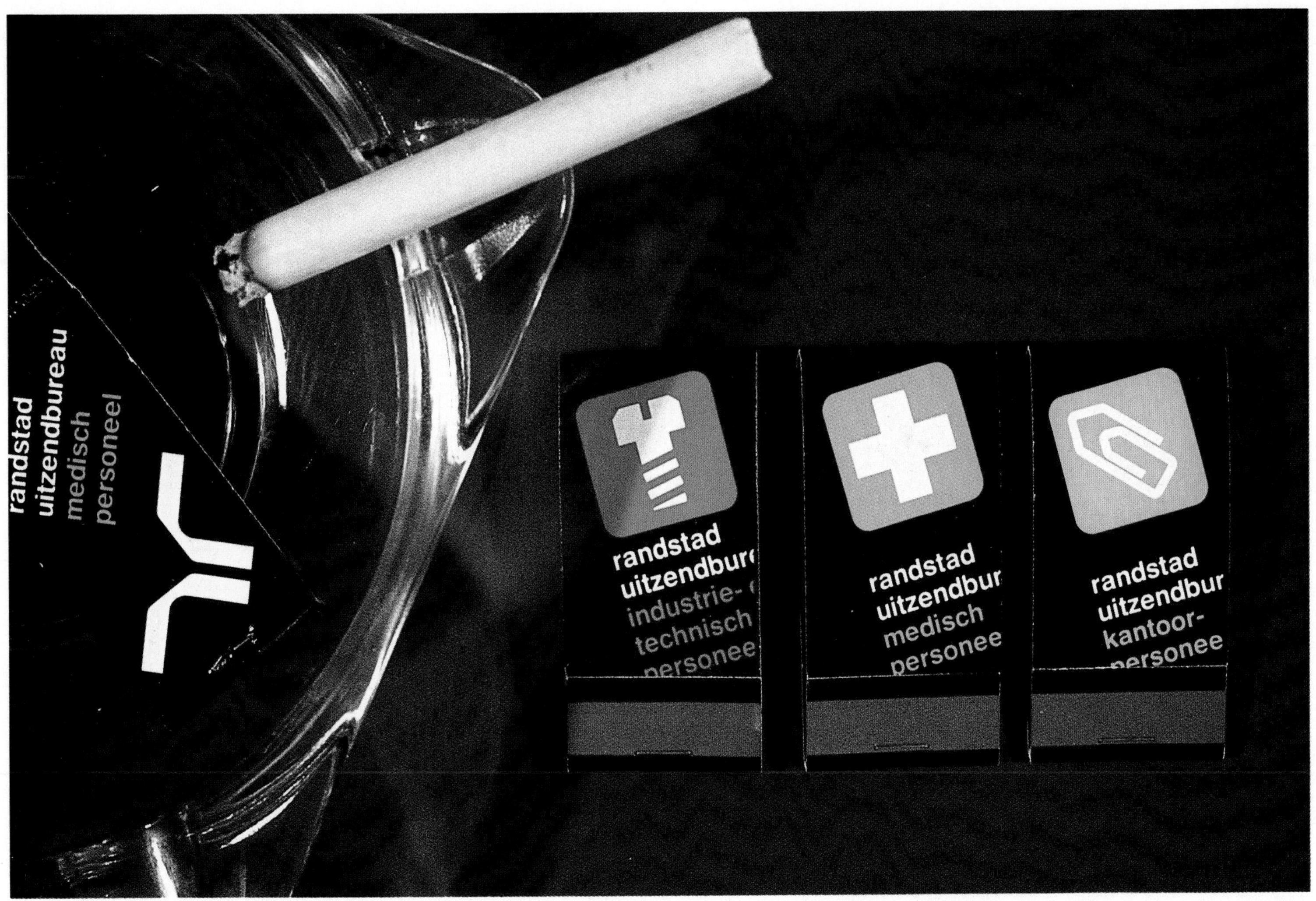

Product: Matchbox
Designer: Wim Verboven
Design Firm: Total Design, Amsterdam
Client: Randstad Uitzendbureau bv

The Matchbox designs feature deep contrasting colors and an array of symbols.

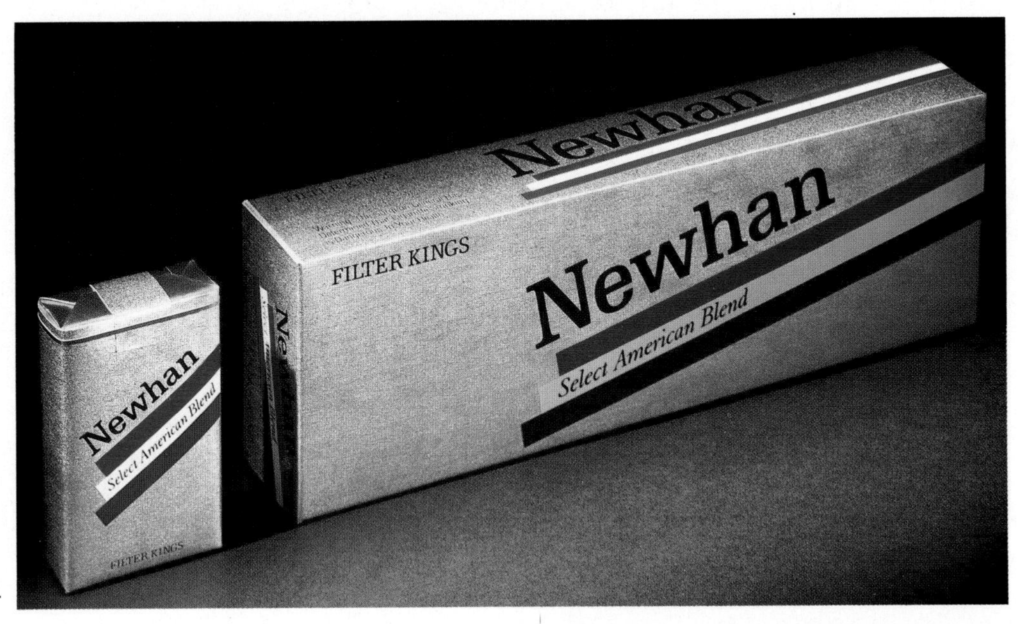

Product: Newhan Cigarettes
Designers: Thomas D'Addario and Dante Calise
Design Firm: The Creative Source Inc., New York, NY
Client: Newhan Allied Corporation

This new product design was geared towards the Far East and Saudi Arabia export markets. The objective was to stress "American" tobacco flavor.

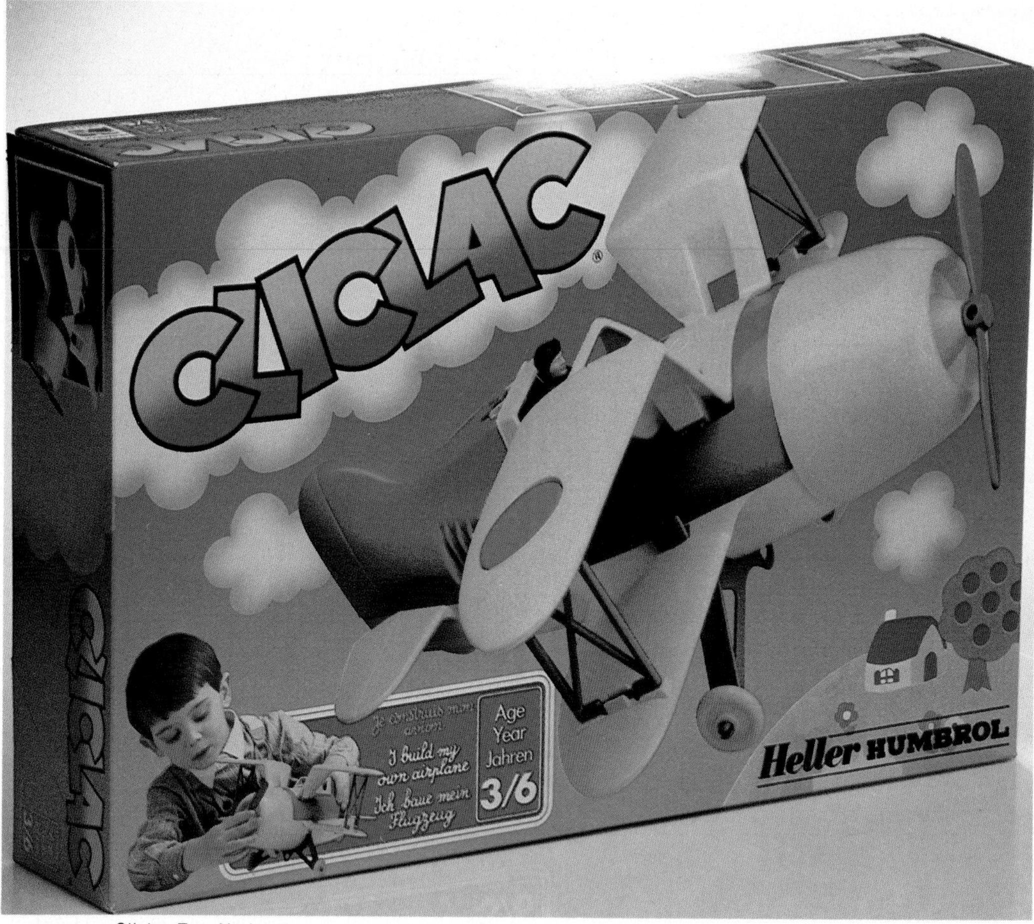

Product: Cliclac Toy Airplane
Designer: P.G.J.
Design Firm: P.G.J., Paris, France
Client: Heller Humbrol

An upscale rendering of this children's toy against an animated background creates an attractive package for the product. The helicopter kit is appropriate for adults.

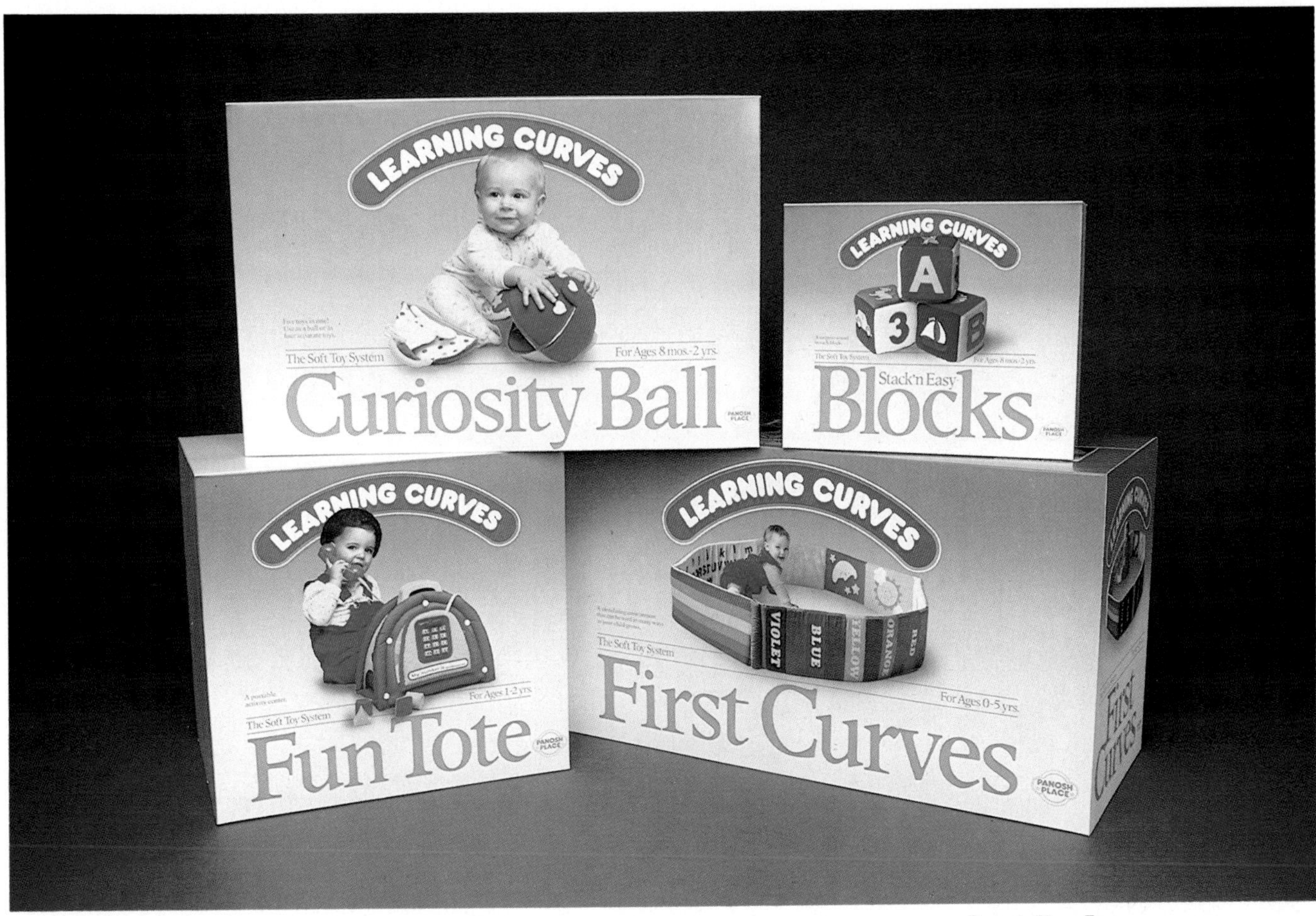

Product: Panosh Place Toys
Designers: Karen Corell and Robin Kupfer
Design Firm: Gerstman & Meyers Inc., New York, NY
Client: Panosh Place, Cherry Hill, NJ

Each package achieves linkage with the umbrella brand by using a distinct logotype. The packaging system is flexible enough to accommodate the addition of new products to the line.

Product: The Transformers
Designers: Edward Morrill and Robert Chapman
Design Firm: Coleman, LiPuma, Segal & Morrill Inc., New York, NY
Client: Hasbro Inc., Pawtucket, RI

Sophisticated air-brushed illustrations, combined with photography and window packaging, bring a high-tech image to these toys.

Product:	Foreign Language Test Book
Designer:	Group Four Design
Design Firm:	Group Four Design, Avon, CT
Client:	McGraw Hill

A unique, contemporary look was given to foreign language test books.

Product: Stockings
Designer: Hotshop
Design Firm: Hotshop, Paris, France
Client: Le Bourget, Paris, France

Le Bourget, a hoisery manufacturer, produces this line of "proportioned pantyhose." Different-colored boxes indicate the various textures and weights.

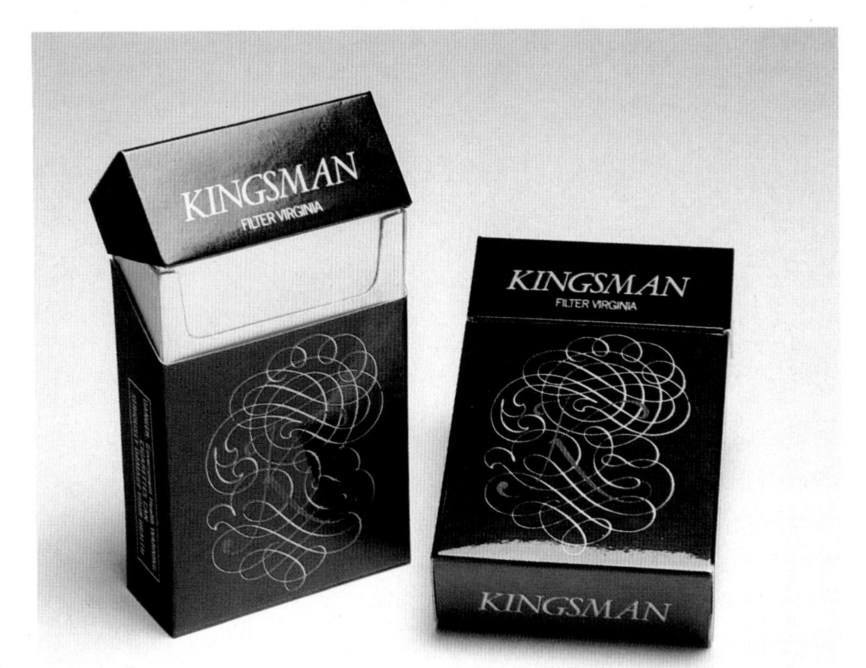

Product: Park Lane Cigarettes
Designers: David Hillman and Nancy Slonims
Design Firm: Pentagram, London, England
Client: Park Lane Tobacco Company

The Kingsmen, a new brand of cigarette launched into a competitive market, is packaged to appear elegant and expensive.

Product: Jewelry Boxes
Designers: Giovanni Brunazzi and Giovanna Ceste
Design Firm: Image and Communication, Torino, Italy
Client: Polidiamants, Italy

These jewelry boxes are made of laquered wood and are lined with black suede.

Product: Pantyhose
Designer: Design Group Italia
Design Firm: Design Group Italia, Milano, Italy
Client: Industra tessile BBB, Italy

These plastic pouches replace the traditional pantyhose packages. The product's color is highly visible through the corners of the envelope.

Product: Tod's Car Shoes
Designer: Massimo Alitti, staff designer
Client: Calzaturificio, Della Valle d. S.p.A., Milano, Italy

This attractive box serves well as a store display unit. The shoes are wrapped in a felt bag that is reusable for travel purposes.

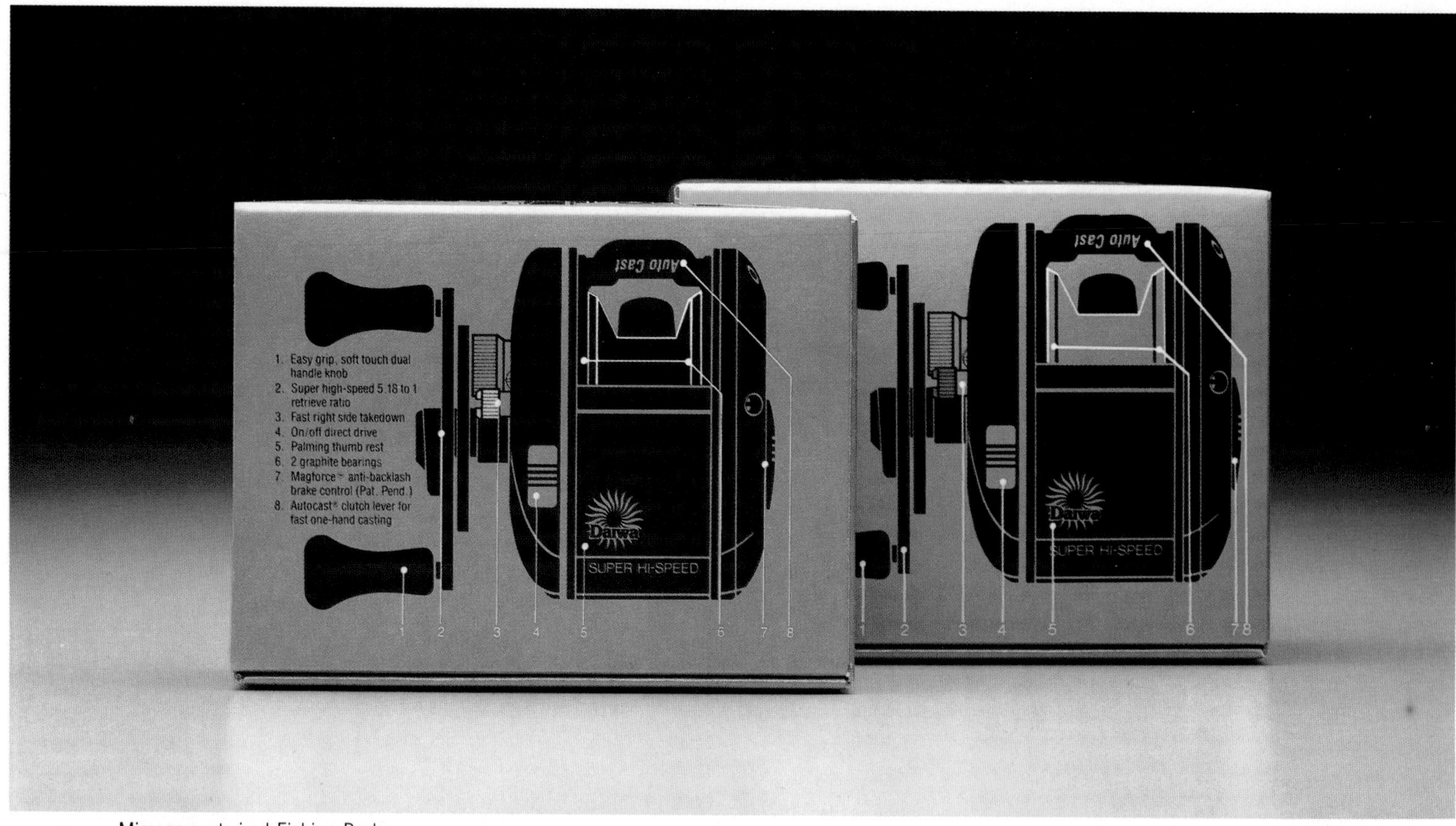

Product: Microcomputerized Fishing Reel
Designers: Julie Devos and Kathleen Campbell
Design Firm: Tom Campbell & Associates Inc., Los Angeles, CA
Client: Daiwa

These fishing reels are packed in self-sell cartons that illustrate and identify the important product features on the panel.

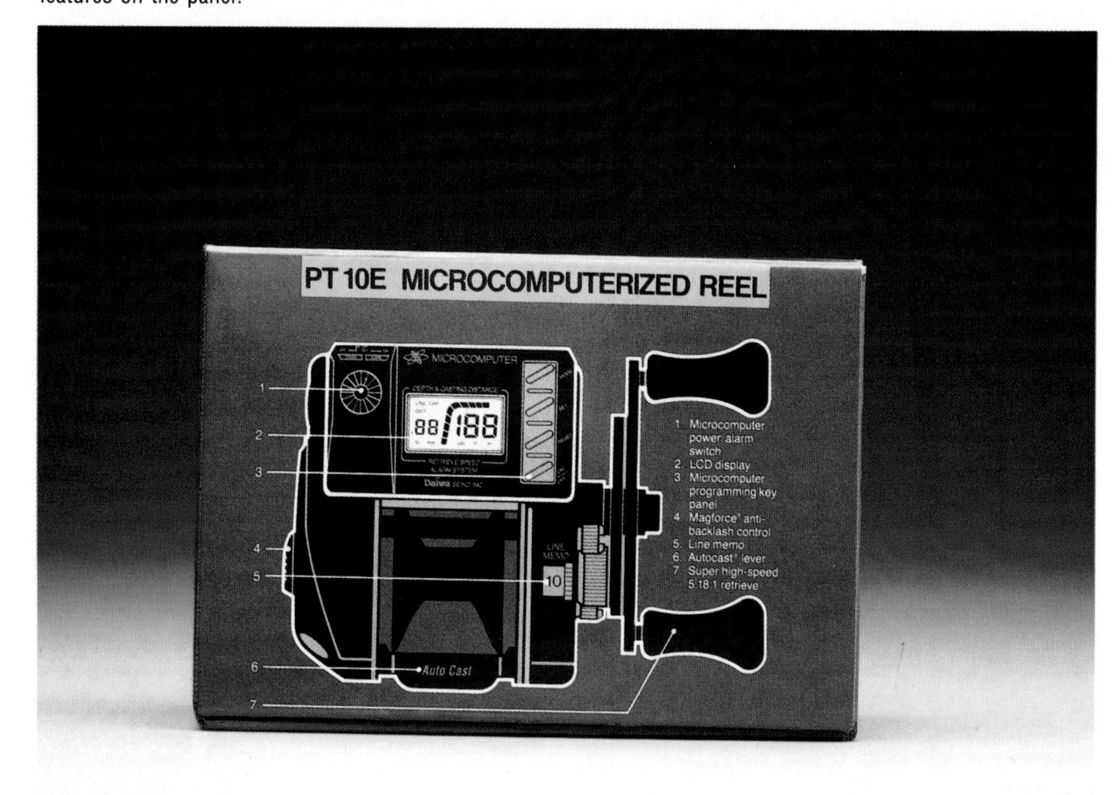

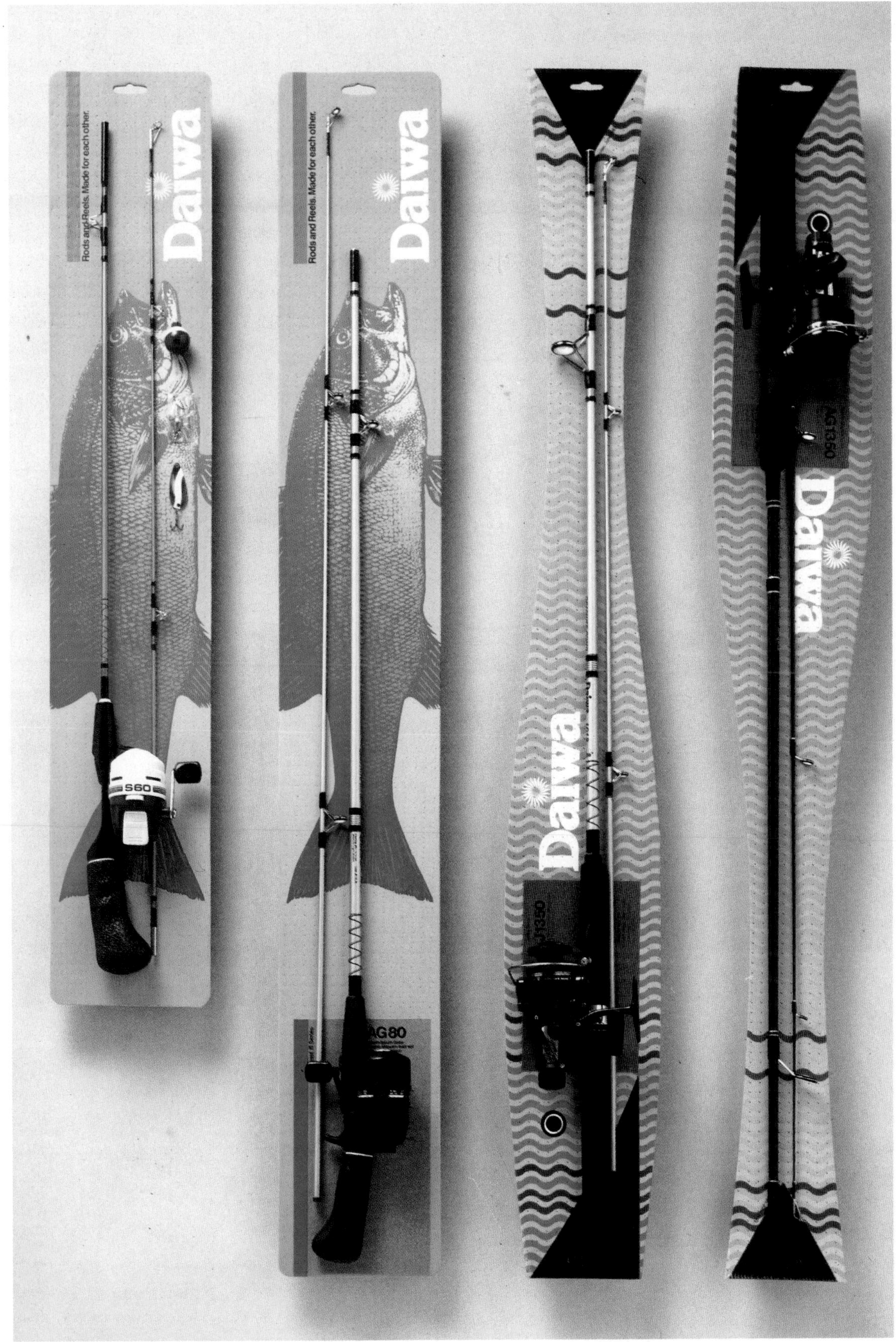

Product: Fishing Rod and Reel Kits
Designers: Kathleen Campbell and Julie Devos
Design Firm: Tom Campbell & Associates Inc., Los Angeles, CA
Client: Daiwa

The skin wrap packages were designed to promote impulse sales of low-cost fishing kits sold in mass-merchandise stores. The large, colorful fish graphics are used to generate a feeling of fun as well as to enhance the display of the product.

Product: YSL
Designer: Wayne Krimston
Design Firm: Murrie White Drummond & Lienhart Associates, Chicago, IL
Client: YSL

A fashionable upscale package for men's underwear reinforces and capitalizes on the YSL designer image.

Product: Hanes Knitwear
Designer: Scott Johnson
Design Firm: Gerstman & Meyers Inc., Chicago, IL
Client: Hanes Knitwear Inc., Winston-Salem, NC

In the redesigning of package graphics for Hanes brand men's and boy's underwear, the objectives were to provide upscale, high-quality imagery, immediate Hanes brand recognition, a masculine brand personality, and a contemporary, uncluttered look.

Product:	Men's Briefs
Designer:	Studio Vu Srl
Design Firm:	Studio Vu Srl, Milano, Italy
Client:	Magnolia S.p.A., Italy

The gray-and-white striped background, together with a product photograph, creates a masculine identity that is quickly identifiable.

Product: Gitano Hosiery
Designer: Marianne Walther
Design Firm: Peterson Blyth Associates, New York, NY
Client: L'Eggs Products, Winston-Salem, NC

This functional design for Gitano Hosiery includes a brightly colored code designating hosiery type and a die-cut window that enables the consumer to feel the texture of the product.

Product: Moathouse
Designer: David Hillman and Lydia Thornley
Design Firm: Pentagram Design Ltd.
Client: Moathouse

Moathouse is a mail order company that promotes Fortnum & Mason products in the United States. The packaging is representative of the fine-quality Fortnum & Mason department stores in London.

Product: Giorgio Armani Products
Designer: Minale, Tattersfield & Partners
Design Firm: Minale, Tattersfield & Partners, London, England
Client: Giorgio Armani, Milano, Italy
Award: 1986 Design and Art Direction award, London

This two-part package, used for various items, employs a sliding system. When closed, the container forms a triangular window, showing the color of the garment inside and the shape of the letter "A".

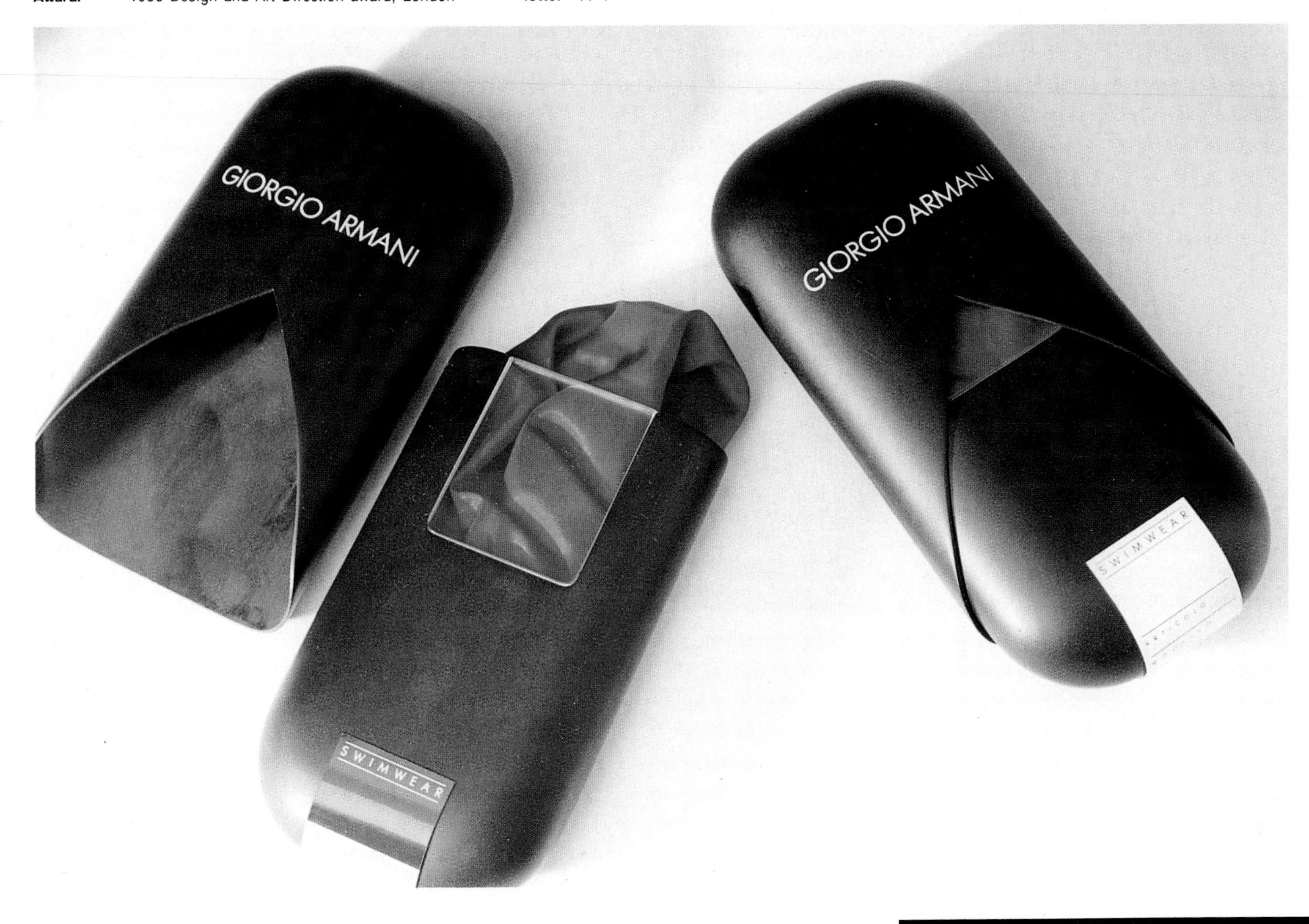

Product: Penn Hi-Speed
Designers: Peter Petronio and Gerald Bot
Design Firm: Concept Groupe, Paris, France
Client: Penn, France

This package was designed for a top-of-the-line tennis ball by Penn. The product name is graphically displayed across the can.

Product: Penn Athletic Products
Designer: Robert Brookson Design
Design Firm: Robert Brookson Design, Phoenix, AZ
Client: Penn Athletic Products, Phoenix, AZ

The objective of Penn Athletic Products was to design a new, high-impact visual merchandising identity, centered on the packaging for Penn's key product lines, while carrying through a unified system of graphics.

Product:	Viyella Cup Collection
Designers:	Keith Murgatroyd, Girvan Roberts, Pippa Thomas, and Leslie Myddleton
Design Firm:	Royle-Murgatroyd Design, London, England
Client:	Viyella, London, England

This design helped launch a new garment line in many retail stores and outlets.

Product: Bookjackets
Designer: John McConnell
Design Firm: Pentagram, London, England
Client: Faber & Faber

These bookjackets, for a series of plays, an art and film series, and the William Golding fictional series, are dramatically designed to identify each series. The play series bookjackets have borders of different bright colors, while the art and film series has a simpler layout. Each volume of the Golding series is enhanced by chilling theatrical illustrations.

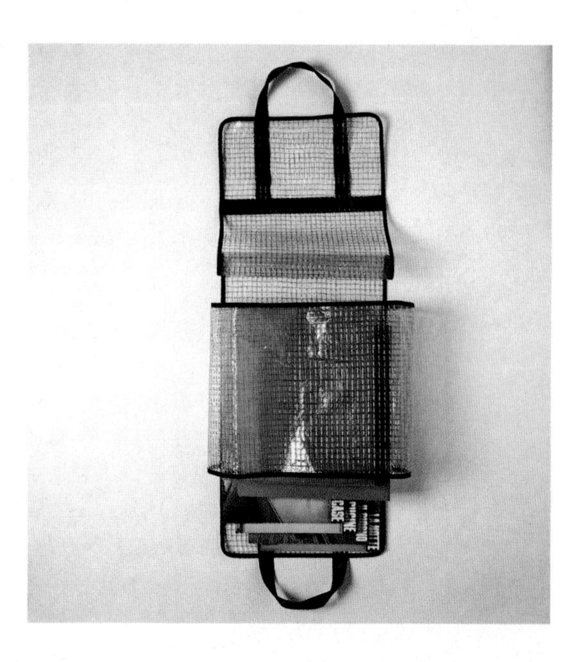

Product: Plastic Carrying Case
Designer: Adelaide Acerbi
Design Firm: Adelaide Acerbi Grafico, Milan, Italy
Client: Driade Stores, Italy

A chic, convenient carrying case for the Driade catalogs is made of transparent plastic so that the name Driade can be seen.

ENDGAME
BY
SAMUEL
BECKETT
SKIRMISHES
BY
CATHERINE
HAYES
A MAP
OF THE WORLD
BY
DAVID HARE
SKINS
OI FOR
ENGLAND
BY
TREVOR
GRIFFITHS

THE LANGUAGE OF MODERN MUSIC
DONALD MITCHELL
HERBERT READ
THE MEANING OF ART
RUDOLF ARNHEIM
FILM AS ART

Product: Veha Handbags
Designer: Adelaide Acerbi
Design Firm: Lambda, Milano, Italy
Client: Veha, Italy

To protect this fine line of handbags by Veha , this packaging system is made of corrugated, heavy paper on the outside and finer, softer paper on the inside. Also on the inside is a soft white ''skarf'' for storage without the box.

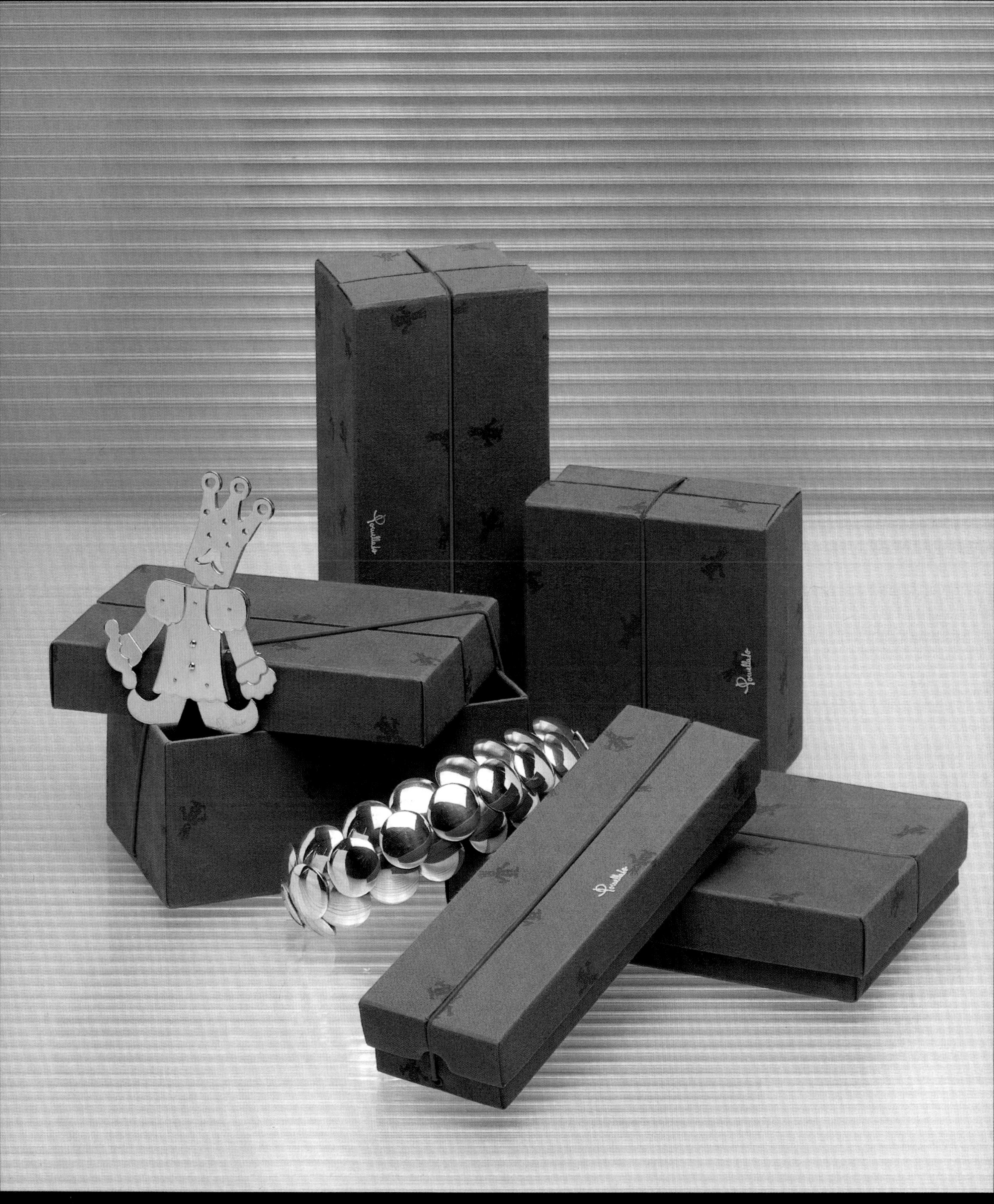

Product: Pomellato Jewelry
Designer: Nabuko Imai
Design Firm: Staff, Pomellato
Client: Pomellato, Milano, Italy

These elegant, canvas-covered boxes serve as durable, easy-to-stack packages for jewelry items.

APPENDIX
International and American Packaging Associations

Argentina

Instituto Argentino del Envasse
Avda. Belgrano 2852
1209—Buenos Aires

Australia

The Australian Institute of Packaging
P.O. Box 20
Chatswood, N.S.W.

Packaging Council of Australia
370 St. Kilda Road
Melbourne, Vic. 3004

Austria

Österreichisches Institut für Verpackungs Wesen
Geschaftselle: Gumpendorferstrasse 6
A-1060 Vienna

Österreichisches Verpackungs—Zentrum
Hoher Markt 3
A-1011 Vienna

Belgium

Institut Belge de l'Emballage
Rue Picard 15
1020 Brussels

Brazil

Associacão Brasileira de Embalagem
Av. Paulista, 688-15°
São Paulo 3

Canada

Packaging Association of Canada
10 St. Mary Street
Toronto, Ontario, M4Y 1P9

Chile

Instituto del Empaque de Chile
Pedro de Valdivia 1481
Santiago

Colombia

Department de Empaques y Embalajes
Proexpo, Apartado Aereo 19766
Bogota, D.E.

Cuba

Comisión Nacional de Envases y Embalajes
Calle 3re No. 3605
Esq. A, 36-A
Miramar, Havana

Czechoslovakia

Imados
U Micheleskeho Lesa 366
Prague

Democratic Republic of Germany

Zentralinstitut für Verpackungs—Wesen
Reisstrasse 42
8017 Dresden

Denmark

Emballageinstituttet
Jemtlandsgade 1
DK 2300 Copenhagen S

England

Alliance Graphique Internationale/AGI
Pentagram
61 North Wharf Road
London W2, United Kingdom

The Design Council
28 Haymarket
London SW1Y 45U

Flexible Packaging Association
31 Craven Street
London WC 2N5NP

The Institute of Packaging
Fountain House, 1A Elm Park
Stanmore, Middlesex HA 74BZ

International Council of Graphic Design Associations/ICOGRADA
12 Blendon Terrace
Plumstead Common
London SE 18 7RS
United Kingdom

Produce Packaging and Marketing Association
15 Hawley Square
Margate, Kent CT9 1PF

Society of Typographic Designers/STD
17 Rochester Square
Camden Road, London NW1, United Kingdom

Egypt

Egyptian Packaging Association —PACKFORICO/EDPA
P.O. Box 2408
Cairo

Federal Republic of Germany

Institut für Lebensmitteltechnologie und Verpackung
Schragenhofstrasse 35
D-8000 Munich 50

Rationalisierungs—Gemeinschaft Verpackung
Postfach 11 91 93
6000 Frankfurt/Main

R G Verpackung im RKW
Gutleutstrasse 163-167, Postfach 11 91 93
6000 Frankfurt/Main

Finland

Finnish Packaging Association
Ritarikatu 3b A
SF-00170 Helsinki 17

France

Centre National de l'Emballage
Avenue Georges Politzer
78—Trappes

Institut Français de l'Emballage
40, rue du Colisée
75008 Paris

Hong Kong

Hong Kong Packaging Council
Eldex Industrial Building, 12/F, 21A Ma Tau
Hung Hom, Kowloon

Hungary

Hungarian Institute of Materials Handling and Packaging
H-1431, P.O. Box 189
Budapest, Rigo u. 3

India

Indian Institute of Packaging
E-2 Marol Industrial Estate, Andheri East
Bombay 400093

Ireland

Irish Packaging Institute
Confederation House, Kildare Street
Dublin 2

Israel

Graphic Designers Association of Israel
P.O. Box 11554
30 Amos Street
Tel Aviv, Israel

The Israel Institute of Packaging
2 Carlebach Street, P.O. Box 20038
Tel Aviv, 61200

Italy

Instituto Italiano Imballaggio
Via Carlo Casan, 34
I-35100 Padova

Jamaica

Jamaica Packaging Association
8 Waterloo Road
Kingston 10

Japan

Japan Packaging Institute
Honshu Building, 12-8 5 Ginza Chuo-ku
Tokyo

Korea (South)

Korea Design and Packaging Centre
128 Yunkun-dong, Chongro-Ku; P.O. Box 23
Seoul

Mexico

Instituto Mexicano de Asistencía a la Industria
Homero 1425-602
Mexico 5, D.F.

Netherlands

Nederlands Verpakkingscentrum
P.O. Box 835
2501 CV, The Hague

New Zealand

New Zealand Institute of Packaging
Box 9130
Wellington
New Zealand

Norway

Den Norske Emballasjeforening
Klingenberg gt 7, Postbox 1754-Vika
Oslo 1

Pakistan

The Packaging Cell of the Export Promotion Bureau
NPT Building
1. 1. Chundriger Road
Karachi Z

Peru

Instituto del Envase y Embalaje del Peru
Las Flores 346
San Isidro, Lima 27

Philippines

Asian Packaging Federation
Far East Building, Room 405, MCC P.O. Box 105
Makati, Rizal, 3117

Poland

Polish Packaging Research and Development Centre
ul. Konstancinska 11
02-942 Warsaw

Portugal

Centro Nacional de Embalagem
Praca das Industrias
Lisbon 3

Romania

Directia Pentru Ambalaje
Calea Victoriei nr 152
Sector 1, Bucharest

Spain

Instituto Español del Envase y Embalaje
Breton de los Herreros, 57
Madrid-3

Sri Lanka

Sri Lanka Institute of Packaging
℅ Aitken Spence Co., LTD
P.O. Box No. 5
Colombo

South Africa

The Institute of Packaging (S.A.)
P.O. Box 3259
Johannesburg 2000

Sweden

Svenska Forpackningsforsknings-institute
Box 91. Hammarby Fabriksvag 29-31
S-12122 Johanneshov 1

Swedish Packaging Institute
Box 9
S-16393 Spanga

Switzerland

Vereingung Schweizerches
Verpackungsinstitut
Verlang Max Binkert & Co.
CH-4335 Laufenburg

Vereinigung Schweiz
Verpackungsinstitut
Dreikonigstrasse 7
8002 Zurich

Taiwan

China Packaging Institute (C.P.I.)
489 Fu-Hsing N. Road
Taipei

Thailand

The Thai Packaging Association
Industrial Service Institute Building, Soi
Rama 4 Road, Bangkok 11

United States

American Institute of Graphic Arts/AIGA
1059 Third Avenue
New York, NY 10021

Association of Industrial Metallizers, Coaters and Laminators (AIMCAL)
61 Blue Ridge Road
Wilton, CT 06897

Clio Awards
336 East 59th Street
New York, NY 10022

Flexible Packaging Association
1090 Vermont Avenue, NW
Washington, D.C. 20005

Glass Packaging Institute
1800 K Street, NW
Washington, D.C. 20006

ID Annual Design Review
Design Publications Inc.
330 West 42nd Street
New York, NY 10036

Industrial Designers Society of America/IDSA
6802 Popular Place
McLean, VA 22101

National Paperbox Association
231 Kings Highway
E. Haddonfield, NJ 08033

Package Designers Council
P.O. Box 3753
Grand Central Station
New York, NY 10017

Paperboard Packaging Council
Suite 600, 1800 K Street, NW
Washington, D.C. 20006

Society of Typographic Arts/STA
Suite 301, 233 East Ontario Street
Chicago, IL 60611

World Packaging Organization/WPO
℅ Packaging Institute of USA
342 Madison Avenue
New York, NY 10017

Young Designers Competition
260 Fifth Avenue
New York, NY 10001

Union of Soviet Socialist Republics

All-Union Scientific Research, Experimental and Design Packaging Institute "WNIEKITU"
Gradcewskoje szose
Kaluga 9

Uruguay

Centro Uruguayo del Empaque
Sarandi 690 D-2° Entrepiso
Montevideo

Venezuela

Camara Venezolana del Envase
2° Piso-B, Esquina de Puente Anauco
Caracas

Yugoslavia

Packaging and Internal Transport Committee
Terazije 23
Belgrade

INDEX I

PRODUCTS

INDEX II
Art Directors and Illustrators

INDEX III
Designers

INDEX IV
DESIGN FIRMS

INDEX V

CLIENTS